Foundations of Epidemiology

Foundations of Epidemiology

MARIT L. BOVBJERG

KELLY JOHNSON

OREGON STATE UNIVERSITY
CORVALLIS, OR

Download for free at https://open.oregonstate.education/epidemiology/

This book was produced with Pressbooks (https://pressbooks.com) and rendered with Prince.

Publication and on-going maintenance of this textbook is possible due to grant support from Oregon State University Ecampus.

Suggest a correction

Contents

Acknowledgements

Many, many thanks to Kelly Johnson for her extensive help with edits and suggestions on earlier drafts of this book. I received further useful feedback from Lindsay Miller, Alicia Bublitz, Leanne Cusack, Kylee Barnes, Julia Drizin, Wafa Hetany, Kindra McQuillan, Colin Mulligan, Michael Murphy, Christopher Skypeck, and Justin Ter Har. I would also like to thank Oregon State Ecampus and the Open Educational Resources unit for their support, including Andrea Fennimore, who helped with layout and formatting, and Daniel Adams, who helped with illustration design.

Preface

Understanding human health—defined by the World Health Organization (WHO) as a "state of complete physical, mental, and social well-being, and not merely the absence of disease or **infirmity**"[i]—is vital. Why are *these* people sick but *those* people aren't? What can we do to improve health for everyone? Improved health in turn leads to gains in economic, social, educational, and other arenas, all of which are necessary for a successful, functioning society.

Unfortunately, to understand human health, we must study humans—and humans are extremely difficult to study. Unlike laboratory-based sciences, where all conditions are under the control of the scientist, conducting scientific studies with human participants includes a host of complications and potential stumbling blocks. First and foremost, humans do not exist in controlled settings like laboratories. Each person has their own job, their own preferred foods, their own sleep schedule, their own hobbies, their own genetics, their own stress levels, their own set of environmental and social settings—all of which affect health. Since no two people are exactly alike, and since we cannot really tell people what to do, studying them is difficult.

Epidemiology is the set of tools we use to study human health. As such, it is not a topic itself *per se*, but rather a set of research methods that are then applied to other health-related topics (kinesiology, infectious disease, cardiology, child development, etc.). Epidemiology can help answer questions such as the following:

- Are dietary-based or exercise-based interventions better for preventing a second heart attack?
- Which people are at the highest risk of dying from influenza?
- Is it safe to eat raw oysters?
- Do birth control pills cause breast cancer?
- How does Zika virus spread?

An understanding of epidemiologic methods is helpful for anyone interested in human health, particularly those in public health or clinical fields. This book is written primarily for use in undergraduate introductory epidemiology classes; however, graduate students, medical students, and professionals working in public or allied health fields may also find it useful either as an introduction or as a refresher.

This book is intended to provide a basic introduction to epidemiologic methods and epidemiologic thinking. After reading this book, you should be able to read an epidemiologic study, understand what the authors did and why, and identify what they found. You will also have the tools to assess the *quality* of that study—how good is the evidence? What are potential sources of bias, and how might those have affected the results? This book will *not* teach you enough to be able to design and conduct your own epidemiologic studies—that level of understanding requires several years of specialized training. However, being able to read and understand the scientific literature about human health will allow you to apply that understanding to your own work in a nuanced, sophisticated way.

It will also allow you to be a confident consumer of the news—how many times have you seen a headline about some new, health-related study and wondered if it could possibly be "real"? Now you will be able to assess for yourself whether the touted new study should change your behavior or not. For instance, for many years we were told that low levels of alcohol consumption, particularly of red wine, were beneficial.[ii] Then during the summer of 2018, the WHO released a statement saying that no, *all* alcohol consumption is harmful.[iii] What should you do? Is a glass of wine a day a good idea or not? This book will provide the tools necessary for you to be able to assess the epidemiologic evidence and decide for yourself.

References

i. Constitution of WHO: principles. *World Health Organization* (WHO). http://www.who.int/about/mission/en/. Accessed October 12, 2018. (↲ Return)

ii. Jaret P. Bottoms up. *WebMD*. https://www.webmd.com/diet/features/health-benefits-wine. Accessed October 12, 2018. (↲ Return)

iii. McKay T. World Health Organization study finds alcohol responsible for five percent of deaths worldwide. *Gizmodo*. https://gizmodo.com/world-health-organization-study-finds-alcohol-responsib-1829247664. Accessed October 12, 2018. (↲ Return)

1. What is Epidemiology?

Learning Objectives

After reading this chapter, you will be able to do the following:

1. Define *epidemiology*
2. Provide examples illustrating each of the 5 parts of the epidemiology definition
3. Describe the way in which epidemiology fits into the overall public health workforce

Public health deals with the well-being of communities, with a focus on disease prevention. This is accomplished "through the organized efforts and informed choices of society, organizations, public and private communities, and individuals."[i] In other words, public health professionals first assess the health status of the **population**, determine the causes of any health problems, design interventions in an attempt to address those problems, and then reassess the population's health to evaluate whether the intervention worked.[ii]

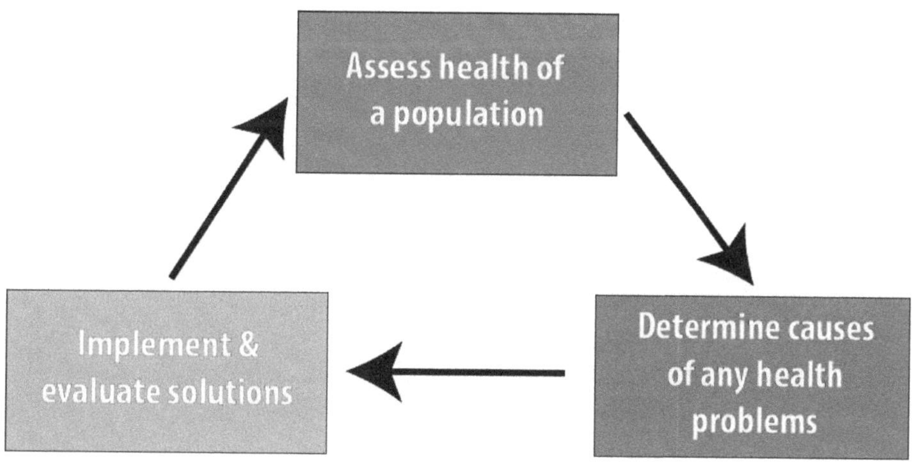

Figure 1-1

Epidemiology is the basic science of public health, and epidemiologists are heavily involved with all 3 steps shown above. We are involved with **surveillance** and other health assessment activities, our studies are instrumental in determining the causes of health outcomes, and we are often part of the teams that evaluate public health interventions.[1]

There exist many closely related definitions of epidemiology. This is the one I like:

> Epidemiology is the study of the distribution and determinants of disease or other health-related outcomes in human populations, and the application of that study to controlling health problems.[ii, iii, iv]

There are several key words and phrases in that definition that relate directly back to the core public health functions. First, epidemiologists are concerned with the *distribution* of a disease — that is, with describing the pattern of an illness in terms of person, place, and time. This **descriptive epidemiology** effort is almost always a necessary first step in any public health initiative.

Disease Is Not Randomly Distributed

Epidemiology as a science works because disease is not randomly distributed within the population. If it were—that is, if there were no risk factors and nothing that would either cause or prevent a case from occurring other than sheer luck—then we would not be able to determine who is at greatest risk. If we could not determine who is at greatest risk, then prevention efforts (the holy grail of public health) would be impossible. However, disease is *not* random. Thus epidemiologists spend their time trying to figure out why *these* people get sick but *those* people don't. Once we figure that out, we have a starting point for planning prevention campaigns with our public health colleagues.

1. Public health interventions comprise any action by a health department, legislative body, or other allied health professional that is designed to improve public health. This could be an education campaign (e.g., billboards about hazards of smoking), surveillance and follow-up, disease outbreak investigation, legislative policy development, outreach efforts (e.g., taking a dental van to elementary schools), etc.

Distribution

Person

Who is getting sick? Men? Women? The elderly? Children? People who live near a particular factory? What do the people who experience the health outcome have in common, and what do the people who do *not* experience the outcome have in common?

As a basic example, let's look at circadian sleep disorder. Based on data published by the WHO, this is the distribution of this condition, broken out by age and sex:

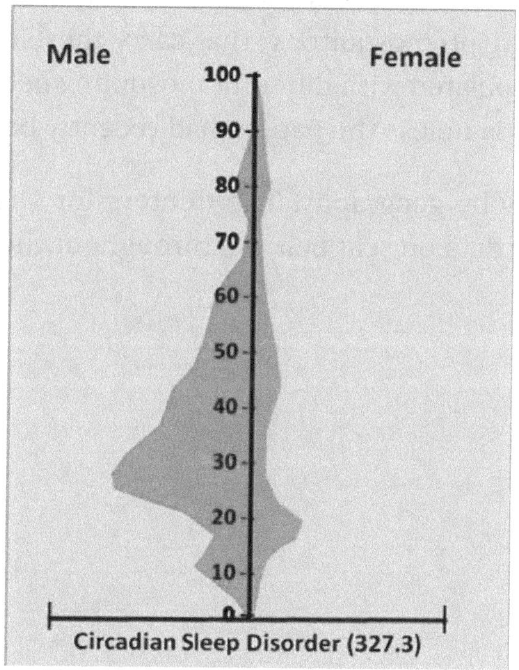

***Figure* 1-2**
Source:
https://brianaltonenmph.com/.../sociocu lturalism-and-health/

Several things become apparent from Figure 1-2. First, circadian sleep disorder is much more common in males than in females. Second, this disorder is most common in adult males in their 20s and 30s. Third, for those females who do have this disorder, it is the most common in adolescents and young adults.

Knowing these patterns—the distribution—of disease would be helpful if you were a clinician trying to diagnose a patient complaining of troubled sleep. Is the patient a 5-year-old female? If so, then it's probably not circadian sleep disorder. Knowing patterns of disease is also useful for public health departments, so they can plan what to do with their limited resources. A health education campaign about healthy sleep habits, according to this figure, would get the most "bang for the buck" if it were targeted at young and middle-aged adult males.

Place

Like other demographic characteristics, where people live has implications for their health. The classic example is a polluting factory: those families who live closer to the contamination may have poorer health than those living farther away or upwind. Infectious diseases can also vary by geographic region—mosquito-borne diseases such as malaria, dengue, and Zika are common in the tropics, where the *Aedes aegypti* mosquitoes[2] that carry these diseases live, and do not occur in colder regions, which are populated with different mosquito species. Thus a physician in Maine would likely not diagnose malaria unless the patient had recently been traveling to the tropics.

Health behaviors can also vary by geography. The Centers for Disease Control and Prevention (CDC) recently published these data on seat belt use throughout the US:

2. Mosquitoes here are the disease *vector*—see http://www.who.int/news-room/fact-sheets/detail/vector-borne-diseases for more information about infectious disease vectors.

Figure 1-3
Source: *https://www.cdc.gov/.../seatbelt_map.html*

Again, studying and understanding the distribution of disease or behavior by place matters for both clinicians and public health professionals: a health department in Minnesota probably does not need to spend resources encouraging seat belt use, whereas in Montana, this might be an excellent use of resources.

Time

The final important factor in descriptive epidemiology is time: How is the distribution of the disease changing over time? For instance, Figure 1-4 gives us a picture of cesarean section rates over 15 years in New South Wales, Australia:

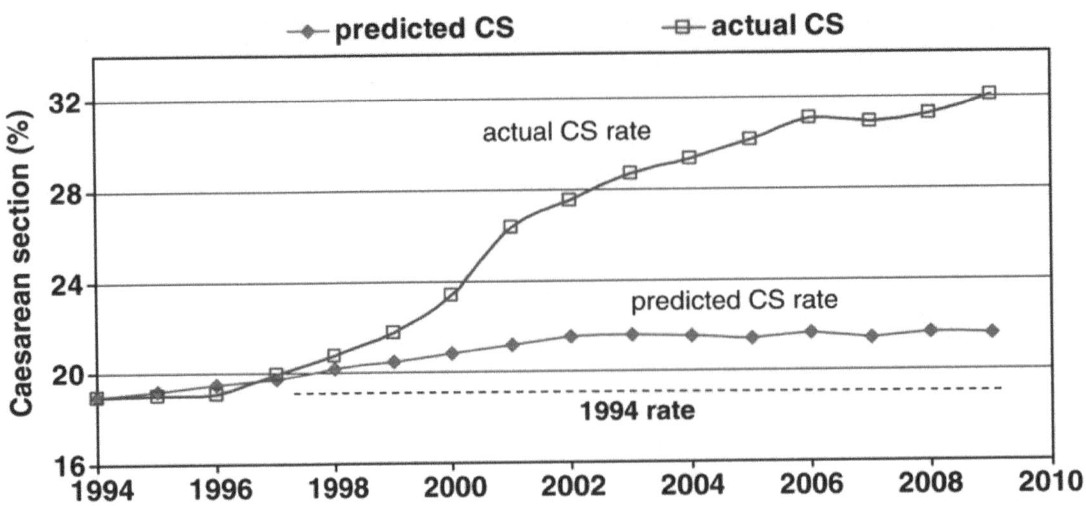

Figure 1-4: *Predicted and actual caesarean section rate between the years of 1994 and 2010.*
Source: *https://bmjopen.bmj.com/.../e001725*

The red line indicates the rate of cesarean sections that would be predicted based on pregnancy risk factors, such as maternal height, maternal blood pressure, and whether it is a multiple pregnancy (e.g., twins). You can see that the cesarean rate based on these known predictors was expected to rise only slightly during the 15 years covered by this graph. However, the blue line shows the actual rate, which rose much more quickly than we would have expected. To the extent that cesarean surgery carries risks (major surgery always has risks), this unexpected jump in cesareans—observed throughout the world, not just in Australia—is alarming[v]. Indeed, efforts to reduce the cesarean rate are currently a top priority for obstetricians, midwives, and public health officials worldwide.[vi, vii, viii, ix]

Determinants

In addition to *who*, epidemiologists are also interested in *why*, which brings us to causes, or **determinants**. We will cover epidemiologic causal theory in depth in chapter 10; here, I will just introduce the topic briefly.

In epidemiology (and throughout this book), when we say "cause," we mean "cause or prevent." In other words, a *cause* (determinant) is anything that changes the likelihood that an individual will become diseased. Sometimes a determinant increases this chance (e.g., smoking); other times, a determinant decreases this chance (e.g., exercise). By this logic, both smoking and exercise are "causes" of disease—the former increases the risk of a variety of conditions, and the latter generally reduces risks. In epidemiology, a determinant, or cause, can be anything that meets the criterion of altering one's risk of disease: behaviors, demographics, genetics, environmental contaminants, and so on. Collectively, all determinants of that disease are called the **etiology** of a disease.

Cause or Disease?

Health behaviors are unique in epidemiology in that they can, depending on context, be both determinants and diseases. For example, smoking causes lung cancer (it's a determinant). However, in an evaluation of a smoking cessation program, smoking is the outcome (it's the "disease"). The same is true for physical activity, nutrition, etc.

Disease

Much like cause, *disease* is an interesting word in epidemiology—it is used to mean "any health-related condition or outcome." Epidemiologists study all manner of health outcomes. Some are "diseases" in the traditional, illness sense: measles, HIV, diabetes, and leukemia. Others are definitely health outcomes but aren't a disease *per se*: pregnancy, malnutrition, physical activity, death. Throughout this book, the word *disease* will be used to refer to any health outcome regardless of whether it is traditionally thought of as a disease in the sense of illness.

Populations

Epidemiologists concern themselves with populations, not individual people. This is both a great asset and a source of great confusion!

First, a definition: a population is a group of people with a common characteristic. This could be residents of the United States, people with type 1 diabetes, people under age 25 who work full-time, and so on. For epidemiologists, the population is the group of people about whom we wish to be able to say something. For instance, say that we are interested in whether the amount of sleep a student gets is related to his or her grade point average (GPA). If we are mainly interested in this relationship among college students, then our population might be "full-time undergraduates." However, there are a lot of full-time undergraduates in the world; we cannot possibly enroll them *all* into our study. We therefore draw a **sample** from the **target population** and do the study with the people in the sample (which here will be some smaller group of full-time undergraduates).

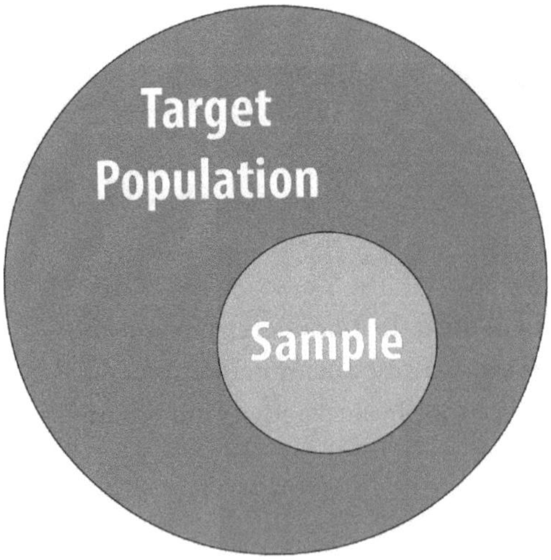

Figure 1-5

Ideally, the sample will be similar enough to the target population that our results can indeed be generalized back to that population (remember, the target population is the group we want to say something about); therefore, we would work to recruit a diverse sample of students who are similar to the population. We would be hard-pressed to generalize to all full time undergraduates if our study was done only among first-year biology majors. However, note that the **generalizability** of our sample does not always matter as much as it does in other fields (see chapter 6 for a lengthier discussion of **external validity**).[x]

Inclusion/Exclusion Criteria

We define populations via lists of inclusion and/or exclusion criteria. These are just flip sides of the same coin: you can either *include* kids or *exclude* adults. It usually doesn't matter whether inclusion or exclusion criteria are used; whichever provides the greatest clarity is generally the best choice. When defining a population, the list of inclusion and exclusion criteria must be sufficiently complete that any given person could look at it and decide whether they were in the population.

As an example, if we were planning a study of strength training to prevent **osteoporotic** fractures in elderly women, the inclusion/exclusion criteria would need to specify the following:

- the lower (and potentially, upper) age cutoff (i.e., what is "elderly" for our purposes?)
- whether we are interested in biological females, those who identify as women, or both
- whether there are any exclusions in terms of physical capabilities (e.g., not all elderly women are able to do a strength training regimen)

We might choose to exclude women for whom exercise of any kind is contraindicated (e.g., if they are heart failure patients), or those who have already had a hip fracture, and so on.

Note that, when creating inclusion/exclusion criteria lists, only rarely is there a "correct" answer. Often, scientific or clinical considerations will help narrow it down, but in our example above, it probably doesn't matter if we set the lower age bound at 60, 65, or 70, as long as we set one and stick to it. Occasionally, policy reasons dictate that one group or another is chosen—for instance, Medicare in the US covers individuals ages 65 and older, so we often see studies in this age group specifically.

Let's reflect back on our definition of epidemiology: we are looking at distributions and determinants of disease in populations. This is an important point to understand—physicians, nurses, and other clinicians are concerned mainly with diseases in individuals, whereas in epidemiology, our focus is instead on populations. This can make interpreting epidemiologic results difficult, since epidemiologic results pertain to populations, not individuals.

For instance, recent data on deaths related to the opioid epidemic suggest that the riskiest state in the US is West Virginia (43.4 opioid-related deaths per 100,000 people in one year) and the least risky is Nebraska (2.4 per 100,000 people in one year).[xi] Note that these statistics say *nothing* about individual levels of risk—all that they say is that, on average, people from West Virginia are much more likely to die from opioid-related causes than people from Nebraska. These are population-level data. For any one individual, we must consider much more than just where the person lives, although that is certainly relevant. A person addicted to painkillers in Nebraska surely has a higher risk of opioid-related death than does a person in West Virginia who has never taken any pain medicine stronger than aspirin, even though the population-level risks might suggest otherwise.

Population-level statistics are a powerful tool because they allow us to do the work of epidemiology—figuring out why some groups (populations) are at higher risk than others. However, when looking at **aggregated** statistics, one must always keep in mind that any one individual's risk is lost within the group. For example, returning to our sleep/GPA study, suppose we are concerned with the average amount of sleep for male and female students. We define our population as full-time, on-campus undergraduates at 4-year institutions in the US (these are the inclusion criteria). We decide to draw our sample from Oregon State University (OSU) and recruit 4,000 students—2,000 males and 2,000 females—to be in our study. We determine that male students sleep an average of 7.2 hours per night and that female students sleep an average of 7.9 hours per night. While this comparison may allow us to comment on differences between male and female students on a population level, it says nothing about individual students. Within our sample, it would be relatively easy to find a given pair of students wherein the male student averaged more sleep than the female.[3] When interpreting epidemiologic data, therefore, it is always important to remember that the data refer to *groups* of people—not to individuals.

Controlling Health Problems

The final piece of our definition is that epidemiology uses all of the information on distributions

3. Also masked within these averages, of course, are personal variations. Even if I average 7.4 hours of sleep per night, certainly there are nights where I get less and nights where I get more. See chapter 6 for a more thorough discussion of this and other issues.

and determinants of diseases in populations *to control health problems*. This final application step is not included in all epidemiology definitions, and indeed whether it should be is controversial in the field.[xii] To my way of thinking, however, the rest does not matter without this step. Epidemiology is the fundamental science of public health, and public health is concerned with preventing disease and improving general wellness in the public. Merely knowing that male students get less sleep than female students does us little good. Who cares? The way we can contribute to the health of the public is by *taking action* based on this knowledge. Imagine if the epidemiologists who first made the link between smoking and lung cancer had not acted on their findings![xiii]

Epidemiologic data are a key part of numerous possible public health actions, including health education campaigns, policy or regulation changes, clinical practice changes, and many other initiatives. Rarely do epidemiologists take this step by themselves—collaboration with professionals from other fields within and related to public health is a must. However, the data generated by epidemiologists are fundamental to planning these actions, the effectiveness of which should always be formally evaluated (a process that often involves epidemiologists) to make sure they worked as intended.

Epidemiology Two Ways

The word *epidemiology* can refer both to the set of methods we use to study the distribution and determinants of disease (as I have used the term thus far in this chapter) – as well as to the collected body of knowledge for a particular health outcome gained as a result of that study. For instance, everything we know thus far about risk factors and **prognoses** for heart failure can collectively be referred to as "the epidemiology of heart failure."

Conclusions

Epidemiology is an important field within public health. Epidemiologists study disease patterns within populations to determine risk profiles and potential health-improvement targets, and they collaborate with others to implement data-driven, population-level, health-related interventions.

References

i. Introduction to public health. Centers for Disease Control and Prevention (CDC). https://www.cdc.gov/publichealth101/public-health.html. Published December 20, 2017. Accessed August 21, 2018. (↲ Return)

ii. Aschengrau A, Seage GRI. *Epidemiology in Public Health*. 3rd ed. Burlington, MA: Jones and Bartlett Learning; 2014. (↲ Return 1) (↲ Return 2)

iii. Porta M. A *Dictionary of Epidemiology*. 5th ed. New York: Oxford University Press; 2008.> (↲ Return)

iv. MacMahon B, Trichopoulos D. *Epidemiology: Principles and Methods*. 2nd ed. Boston: Little, Brown; 1996. (↲ Return)

v. Lancet T. Stemming the global caesarean section epidemic. *The Lancet*. 2018;392(10155):1279. doi:10.1016/S0140-6736(18)32394-8 (↲ Return)

vi. American College of Obstetricians and Gynecologists. ACOG practice bulletin no. 115: vaginal birth after previous cesarean delivery. *Obstet Gynecol*. 2010;116(2 pt 1):450-463. doi:10.1097/AOG.0b013e3181eeb251 (↲ Return)

vii. Reducing primary cesareans. American College of Nurse-Midwives. 2018. http://birthtools.org/Reducing-Primary-Cesareans-NEW. Accessed October 12, 2018. (↲ Return)

viii. U.S. birth data underscores higher C-section risks, CDC says. *Reuters*. 2015. https://www.reuters.com/article/us-usa-health-cesarean/u-s-birth-data-underscores-higher-c-section-risks-cdc-says-idUSKBN0O524X20150520. Accessed October 12, 2018. (↲ Return)

ix. WHO statement on caesarean section rates. World Health Organization (WHO). http://www.who.int/reproductivehealth/publications/maternal_perinatal_health/cs-statement/en/. Accessed October 12, 2018. (↲ Return)

x. Rothman KJ, Gallacher JEJ, Hatch EE. Why representativeness should be avoided. *Int J Epidemiol*. 2013;42(4):1012-1014. doi:10.1093/ije/dys223 (↲ Return)

xi. Abuse NI on D. Opioid summaries by state. National Institute on Drug Abuse. 2018. https://www.drugabuse.gov/drugs-abuse/opioids/opioid-summaries-by-state. Accessed September 7, 2018. (↲ Return)

xii. Keyes K, Galea S. What matters most: quantifying an epidemiology of consequence. *Ann Epidemiol.* 2015;25(5):305-311. doi:10.1016/j.annepidem.2015.01.016 (↲ Return)

xiii. Doll R, Hill AB. Smoking and carcinoma of the lung. *Br Med J.* 1950;2(4682):739-748. (↲ Return)

2. Measures of Disease Frequency

Learning Objectives

After reading this chapter, you will be able to do the following:

1. Define and calculate prevalence
2. Classify individuals as either at risk of disease or not
3. Define and calculate incidence proportion
4. Construct intervals of person-time at risk for a given population
5. Define and calculate incidence rate
6. Differentiate between incidence and prevalence, and explain the mathematical relationship between them

In public health, we often want to quantify disease—how many people are affected by this health outcome? At first glance, this might seem like a simple question, but once you consider the many applications of quantifying disease, the complexities become apparent. In this chapter, you will learn about 3 **measures of disease frequency**: **counts**, **prevalence**, and **incidence**.

Counts (a.k.a. Frequencies)

Sometimes, particularly for extremely rare conditions, we only need to know how many people are sick. How many cases of disease X or health behavior Y were there? A count is just a number—there are no fractions, numerators, or denominators, and the units are always "people."

During the 2017/2018 academic year, for instance, an outbreak of meningococcal meningitis at Oregon State University (OSU) was quantified by *counts*: 6 students got sick.[i]

From **surveillance** data (see chapter 3), we know that the expected number of cases of meningococcal meningitis in a given year is zero. Therefore, 6 cases constitute a level quite above what is expected and would be termed an **epidemic** (see chapter 3).

For rare conditions like this one, simply knowing *how many* cases there are is sufficient for a proper public health response. Since we normally expect none, officials at OSU and the local health department just needed to know that there were 6 students with meningococcal meningitis in order to mount a response (in this case, requiring students 25 years old and younger to be vaccinated).[i]

Similarly, we could look at animal rabies cases observed in Oregon over a 10-year period:

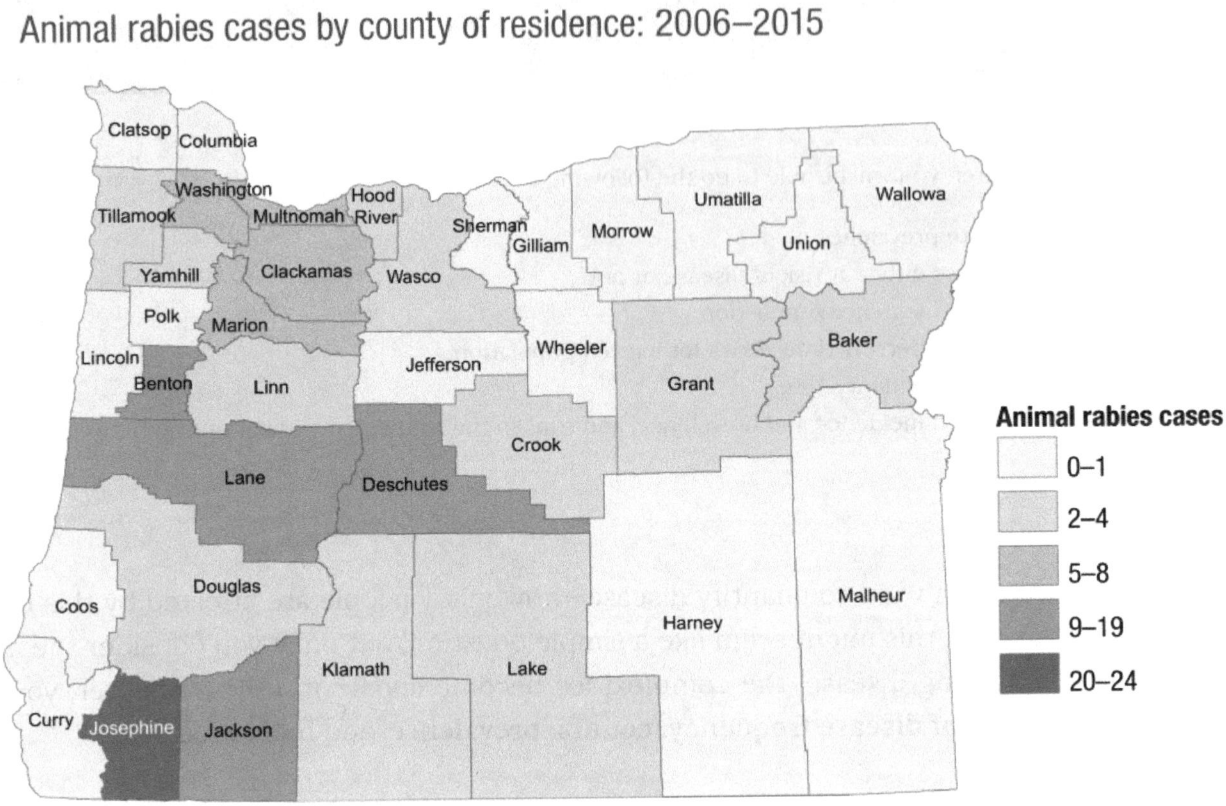

Figure 2-1

Source: *https://www.oregon.gov/oha/PH/.../2015-Rabies.pdf*

The above picture is useful from a public health infrastructure perspective[1]; if you work at the health department in Josephine County, then you might want to keep a few doses of rabies vaccine on hand (since cases of rabies in animals are often discovered because the infected animal bit a human, who must then be vaccinated). However, if you work at the health department in Wallowa County, which has had no recorded cases of animal rabies in 10 years, then maybe your resources would be better spent on things other than vaccine doses that will likely expire before they are

1. The final line in the 'Animal rabies cases' key has been edited to read '20-24.' The original source image reads '2-24.'

used (assuming you could quickly get doses of the vaccine from the state or neighboring counties if they ever became necessary).

Counts are less useful if we want to compare 2 populations. For instance, 1,000 cases of flu in Ashland, New Hampshire, versus 100,000 cases of flu in New York City—we cannot compare these 2 figures at a glance, because the denominators (i.e., the number of people living in each city) are so different.

Incidence and Prevalence

There are 2 commonly used measures of disease frequency that incorporate denominator information: one is a measure of existing disease (*prevalence*), and the other is a measure of new disease (*incidence*). Incidence is used to study causes of disease, whereas prevalence is used more for resource allocation.

Prevalence

Prevalence is a proportion, meaning that everyone who appears in the numerator must also appear in the denominator. This also means that prevalence ranges from zero (no one has the disease) to one (everyone has the disease), and it is usually expressed as a percent.

Prevalence gives us a snapshot of the population-level disease burden at a given time. The formula for prevalence is

$$\frac{\text{\# cases present in the population at a specified time}}{\text{\# people in the population at that time}}$$

Looking at the formula for prevalence, you can see that everyone in the numerator is also in the denominator. Like counts, prevalence is used for resource-planning purposes. Consider the following question a public health authority might be faced with: How much money should our county health department spend on health education about smoking versus on physical activity? One metric for deciding might be which behavior is more *prevalent* in the local population.[2]

The "at a specified time" part of the prevalence definition could refer either to a specific date (e.g., what was the prevalence of flu in Newport, OR on January 22, 2018?) or to a time point in people's lives (e.g., what is the prevalence of breastfeeding at 6 weeks **postpartum**?).

The numerator for prevalence is *all current cases*. Whether the cases were diagnosed yesterday or 20 years ago doesn't matter; both would appear in the numerator. Thus prevalence is affected both by the rate at which new cases occur (the *incidence*, see below) and by how long people typically live with disease. Prevalence is therefore less useful for conditions such as a cold or the flu (where people recover quickly) because once they recover, they are no longer a prevalent case. At any given point in time, most people *don't* have the flu, and so it would seem like the disease burden is quite low based on **point prevalence**, or prevalence calculated on a specific date. In such instances, we sometimes calculate **period prevalence** instead, which is just prevalence of disease over the course of a longer time frame: for example, what was the prevalence of flu in Newport, OR during the entire 2017-2018 flu season? The numerator here would be all of the cases that occurred at any time during those months (counting only the first instance if anyone was unlucky enough to have influenza twice), and the denominator would be everyone who lived in Newport during those same months.

Fig. 2-2

As another example, in Figure 2-2, the prevalence of being light orange is 4/12 = 0.33 = 33%. Note

2. We would use counts to make these decisions when the condition in question is extremely rare, such that the prevalence would be 0.000001% or similar. Then denominators are not as important—this is the case for rabies, as discussed above.

that prevalence does not have units (though providing the *specified time* is often appropriate and never wrong).

Prevalence Examples

Example 1 (Hypothetical Data)

If there are 5,000 students who live in the dorms at Oregon State University (OSU), and during winter term 2018, 400 of them had the flu at some point, then the prevalence of flu was

> 400/5,000 = 0.08 = **8.0% of students living in the dorms during winter term 2018 had the flu at some point**

The above is an example of a *period prevalence*, since we were calculating it over a time period longer than one day. It is also an example of the *specified time* being calendar time—for everyone involved, the specified time was the 2018 winter term.

Example 2 (Based on Known Birth, Infant Death, and Breastfeeding Rates in Oregon)

In 2012, 48,972 babies were born in Oregon. At 14 weeks postpartum, 33,399 of them were being breastfed, and 146 had died. What is the prevalence of breastfeeding at 14 weeks postpartum?

Here we need to subtract the 146 infants who died before 14 weeks from the denominator, as they are no longer part of the population:

> 33,399/(48,972-146) = 0.684 = **68.4% of infants born in Oregon in 2012 were being breastfed at 14 weeks postpartum**

The above is an example of the *specified time* being a particular point in someone's life: the day on which a given baby turns 14 weeks old varies depending on the day he or she was born. This is not a period prevalence, because everyone was assessed on one day—we have just spread those days out throughout the year.

Example 3 (Based on National Estimates)

You can also reverse the calculations to establish the number of people with a disease, given the prevalence and population size. In a report on bone health by the Centers for Disease Control and Prevention (CDC),[v] the authors reported that the prevalence of osteoporosis among men aged 65 and older was 5.6%, and the prevalence among women aged 65 and older was 24.8%. According to data from the US Census Bureau,[vi] as of July 1, 2017, there were an estimated 22,564,684 men and 28,293,995 women aged 65 and older living in the US. Applying the prevalence, we can estimate that:

22,564,684 × 0.056 = **1,263,622 men aged 65 or older**

and

28,293,995 × 0.248 = **7,016,911 women aged 65 or older**

currently have osteoporosis in the US.

Incidence

Incidence is a tricky word in epidemiology, because while it is always a measure of new cases, there are 2 possible denominators and at least a half-dozen words that all refer to this same thing. Yikes!

The numerator for incidence is *always* the number of new cases of a disease observed over some time period. This means that, to study incidence, you must (1) follow people for some length of time (the length varies according to the disease—a few hours or days for a foodborne illness versus a few decades for some cancers) and (2) start with a **population at risk**—that is, people who are at risk of developing the disease (at risk of becoming a case). Usually, at a minimum, we therefore exclude people who already have the disease—such people cannot become an *incident* case because they are already a *prevalent* case. We also exclude anyone not capable of getting the disease, either because they are immune or because they lack the proper organs (e.g., biological females cannot get testicular cancer). Furthermore, because you are establishing the number of new cases, it is always necessary to include time-based units when reporting an incidence.

One way of calculating incidence is to include in the denominator the number of people who were at risk of getting the condition at the start of your follow-up time period. This calculation yields the **incidence proportion**. It's also called, depending on which source you're reading, the **cumulative incidence**, or the **risk**.[vii]

$$\text{Incidence Proportion} = \frac{\text{\# new cases observed during some time period}}{\text{\# people at risk at the start of the time period}}$$

The incidence proportion is interpreted as the average risk (chance) of developing the disease over some time period.

Incidence examples

Example 1 (Hypothetical Data)

If there are 25 students in a particular class, and one person came to class on Monday of the first week already sick with the flu (this person is a prevalent case—they are already sick, so are not at risk), and 2 more people got the flu on Wednesday of that same week, then what was the incidence of flu during Week 1?

Our numerator would be the number of new cases of flu—here, 2. The denominator is the population at risk, so we must subtract out the student who already has the flu because they are not at risk. So the denominator is 24.

The incidence of flu in that class during Week 1 was thus:

2/24 = 0.083 = ***8.3 per 100 per week***

Although prevalence is usually expressed as a percent, for incidence we use "per 100," "per 1,000," "per 10,000," and so on. The precise power of 10 is not standardized; just choose one that gives you a whole(ish) number of people: 8.3 per 100 is the same as 83 per 1,000.

Additionally, it is vital that you specify the time period over which you observed incidence, because interpretation (e.g., how much of a problem this particular disease is) varies widely depending on time. For instance, 2 cases of breast cancer per 100 women in one week are very different than 2 cases per 100 women in 20 years. The former might warrant public health intervention, while the latter almost certainly would not.

Example 2 (Hypothetical Data)

If we want to compare the incidences between 2 populations, it is important to express them in the same power of 10 (e.g., both must be "per 100" or "per 1,000") and also to convert them to cover the same time frame.[3]

If City A has an incidence of norovirus of 25/1,000 per month and City B has an incidence of norovirus of 500/

3. We don't need to worry about population size, because the incidence calculation accounts for those denominators.

10,000 per year, we cannot compare them. We must first convert one of the denominators and one of the time frames so that they are comparable.

Here we'll convert the numbers for City A. First, the denominator needs to be 10,000, not 1000. If we multiply the incidence for City A by 10/10,[4] the incidence in City A is now 250/10,000 per month.

Then we need to adjust the time frame—here by multiplying by 12 (as there are 12 months in a year). The incidence in City A then becomes:

250 × 12 = **3000/10,000 *per year***

Compared to City B's incidence of 500/10,000 per year, the incidence is higher in City A.

Some of you will have spotted a potentially questionable assumption made above: that the incidence in City A—which was measured only over one month—is constant for the entire year. This may or may not actually be true, and in real life, you would have to do a little digging to determine whether it was likely true or not before declaring that norovirus was more common in City A. What if the 1-month data point was from an anomalous month, when City A had a huge norovirus outbreak, for instance? This is not uncommon, because like flu, norovirus is seasonal.

We have thus far been looking at incidences with a relatively straightforward population calculation—for example, the number of students living in a particular dorm at a particular time. The other kind of incidence is the **incidence rate**. Some epidemiology texts will call this the **incidence density**.[vii] Importantly, the numerator is still the number of new cases observed over a given period of time. But the denominator is now the sum of the **person-time at risk**.

The need for this "other" kind of incidence stems from the fact that populations are not static: some people are born, others die, people move in and out. Thus if you quantify the population at risk at the start of your observation window, you are at best only approximating the population, particularly if you follow people for a year or more. Instead, we could look at each person in the population and determine how long they were at risk. Figure 2-3 shows a hypothetical population with 10 people, all of whom were at risk at the start of a 1-year follow-up observation period:

4. Recall from your Algebra classes that this is a legal maneuver, because 10/10 = 1, and you can multiply numbers by 1 with impunity.

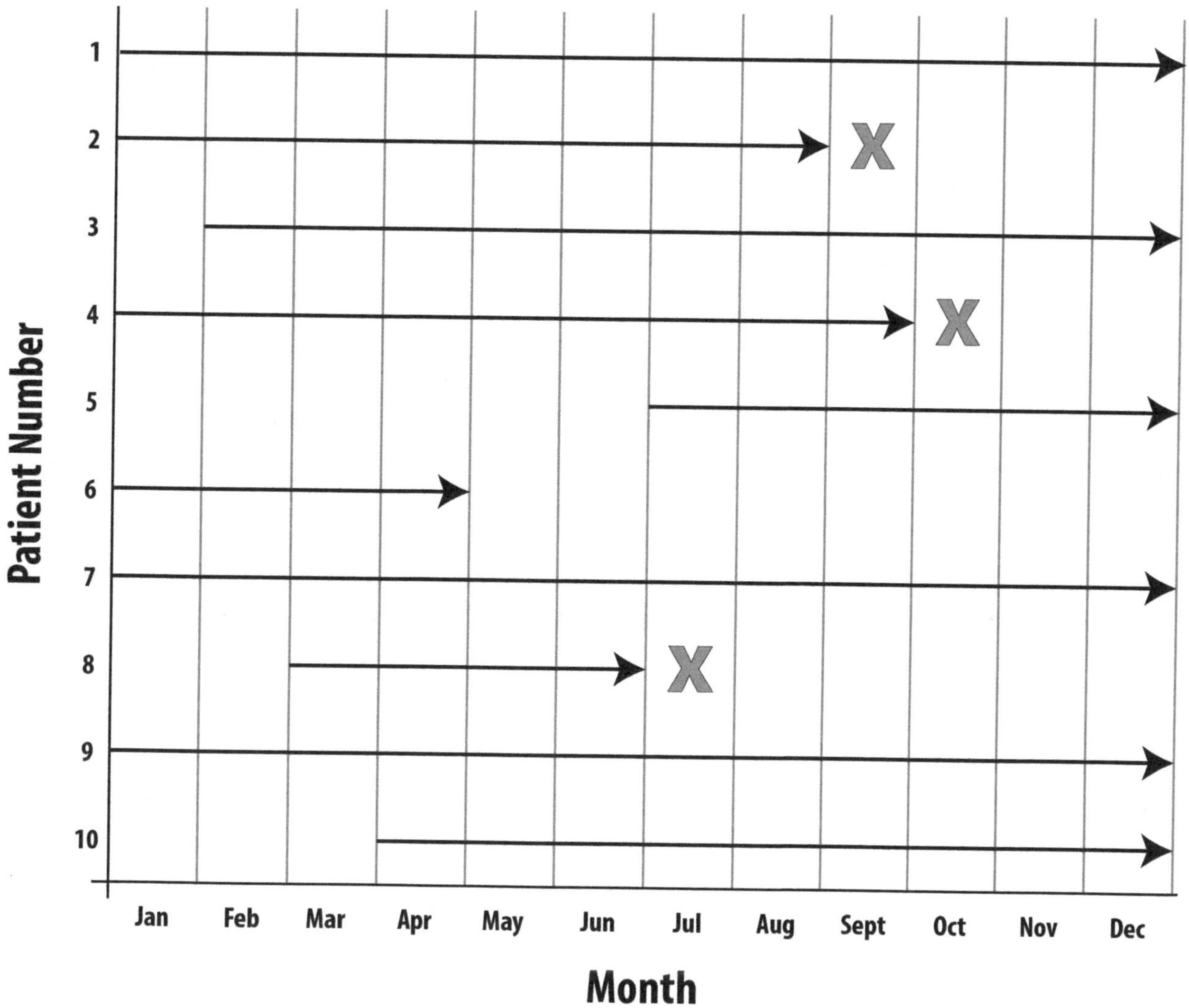

Figure 2-3

Person 1 enrolled in the study January 1 and was followed through December 31 without developing the disease of interest (they may have been diagnosed with other things, but if they have not contracted the disease we're studying, then they're still at risk). Person 1 contributed 12 person-months at risk.

Person 2 enrolled January 1 and developed the disease at the end of August. Person 2 contributed 8 person-months at risk. Person 2 is still alive after August but can no longer contribute person-time *at risk* because now they are a prevalent case.

Person 3 didn't enroll until February 1 and was then followed for the rest of the year without developing the disease of interest. Person 3 contributed 11 person-months at risk.

Person 4 enrolled January 1 and developed the disease at the end of September. Person 4 contributed 9 person-months at risk.

Person 5 enrolled July 1 and did not develop the disease during follow-up. Person 5 contributed 6 person-months at risk.

Person 6 enrolled January 1 and was lost to follow-up at the end of April (this person could have moved away, stopped returning calls, or maybe died of something else—these are called **competing risks**). Person 6 contributed 4 person-months at risk. We can still count these months, because during that time, Person 6 was in the study, and was still at risk—not knowing the outcome (if they moved, etc.) or having their follow-up terminated because of death from a competing risk does not negate the fact that we observed Person 6 for 4 months while they were at risk.

Person 7 enrolled January 1 and was followed through December 31 without developing the disease under investigation. Person 7 contributed 12 person-months at risk.

Person 8 enrolled March 1 and developed the disease at the end of June. Person 8 contributed 4 person-months at risk.

Person 9 enrolled in the study January 1 and was followed through December 31 without incident. Person 9 contributed 12 person-months at risk.

Person 10 enrolled in the study April 1 and was followed through December 31 without incident. Person 10 contributed 9 person-months at risk.

To calculate the incidence rate, then, our numerator is still the number of new cases we observed during the follow-up time—here, there were 3 *new cases* (persons 2, 4, and 8). The denominator is now the sum, in months, of the person-time at risk contributed by all participants.

Calculating Incidence Rate from Data in Figure 2-3

First sum the total person time at risk:

12 + 8 + 11 + 9 + 6 + 4 + 12 + 4 + 12 + 9 = 87 *person-months at risk* (PMAR)

Then calculate the incidence rate:

3/87 PMAR = 0.0345 *per person-month* (PM)

That looks a little ugly, so let's move the decimal place:

3.45 *per 100 person-months*

We could instead express this in terms of years by multiplying our original by 12 (because there are 12 months in the year):

$$(0.0345 \text{ per PM})(12) = 0.414 \text{ per person-year}$$

Finally, we could make it have at least one whole person:

4.14 per 10 person-years

In other words, in Figure 2-3, we observed 4.14 new cases of disease for every observed 10 person-years at risk.

The strengths of the person-time approach are that it allows a more nuanced view of the population at risk and is more realistic: not everyone enrolls in a study on exactly day one, some people experience competing risks or are lost to follow-up, and sometimes a case pops up almost immediately, so that person contributes very little person-time to the denominator (whereas with incidence proportion, they would add a full person).

Limitations of the person-time approach are that it is more complex, and it does not distinguish between 100 people followed for one month (totaling 100 PM), 10 people followed for 10 months (totaling 100 PM), and one person followed for 100 months (still totaling 100 PM). Additionally, loss to follow-up is probably not random (this could also affect incidence proportion if people drop out because they're feeling poorly but before they are recorded as an incident case). It is thus useful to state the time-period over which people were eligible to be followed (in Figure 2-3, one year). Table 2-1 compares the two types of incidence.

Table 2-1: Comparison of incidence proportion and incidence rate

	Incidence Proportion	**Incidence Rate**
Numerator	new cases over a period of time	new cases over a period of time
Denominator	number of people at risk at the start	sum of person-time at risk
You must:	define the time frame	report the person-time units
A.K.A.	risk cumulative incidence absolute risk	incidence density
Range	0-1 (it's a proportion)[5]	0 to infinity

Prior Person-Time at Risk

What about all that time before our study started? If we enroll a bunch of 50 year-olds who are at risk of heart disease and follow them recording the person-time, why not *also* count the person-time at risk from prior to study entry? Each person would yield an extra 50 years of person-time at risk!

We can't do this because we are missing all of the prevalent cases. Some proportion of people develop heart disease prior to age 50, and would thus not be eligible for our study. Without data on how many person-years at risk *those* people had prior to developing heart disease, our incidence would be artificially low, because we add 50 person-years at risk per person to the denominator without accounting for the entire population, which includes some cases that are prevalent by age 50.

Uses of Incidence and Prevalence

As stated above, incidence is used to study the causes of disease. Prevalence is less useful for this because the disease has already happened; we thus have no way of knowing whether the disease or the exposure happened first (necessary for establishing causality). For instance, obesity is associated with lower levels of physical activity—one possible scenario is that lower levels of physical activity lead to obesity, secondary to an energy imbalance. However, another equally possible scenario is that obesity came first and the person subsequently reduced their amount of physical activity, possibly secondary to joint pain. Studying prevalent obesity cases does not allow us to distinguish between these scenarios.

Incidence, on the other hand, can easily be used to study potential causes of disease. When studying incidence, we know that everyone is disease-free at baseline, since we study only the population at risk. Therefore, any exposures assessed at the beginning came before disease onset by definition.

Prevalence is more useful as a way of assessing the disease burden in a particular community, perhaps for purposes of resource allocation. For instance, state health departments in the Northeast and upper-Midwest spend a portion of their budgets on **Lyme disease** prevention

5. if the numerator is the number of cases of a disease that you can get more than once, then incidence proportion can be above 1, because the denominator is still people. For this reason, usually we count only the first episode of a given disease.

education (e.g., billboards about tucking your pants into your socks) because Lyme disease is quite prevalent in those regions:

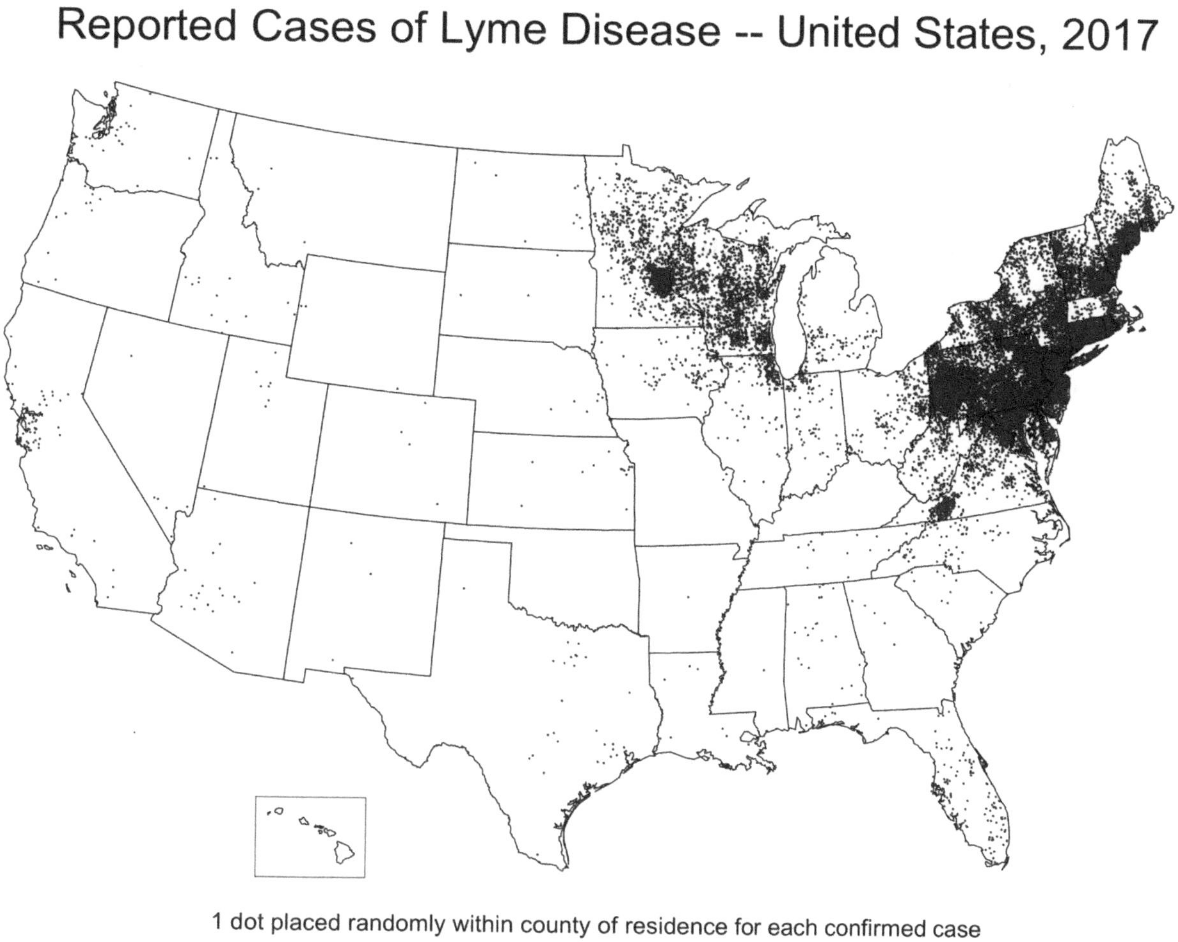

Figure 2-4
Source: *https://www.cdc.gov/.../maps.html*

However, the prevalence of Lyme disease in Colorado is extremely low; health departments in Colorado would do well to spend their money elsewhere. Prevalence data are also useful for health care administrators: if you know that 80% of your nursing home residents have dementia in some form, then this has implications for staffing, standard operating procedures, and potentially even for the layout and design of the space (pictorial signs on the walls to indicate the purposes of rooms, for instance).

Relationship between Incidence and Prevalence

As mentioned above, prevalence is affected by both the incidence (how many new cases pop up) and the disease's duration. If people live longer with a disease, then they remain prevalent cases for longer. Thus

$$\textbf{Prevalence} \approx \textbf{Incidence} \times \textbf{average duration.}$$

Here is an example:

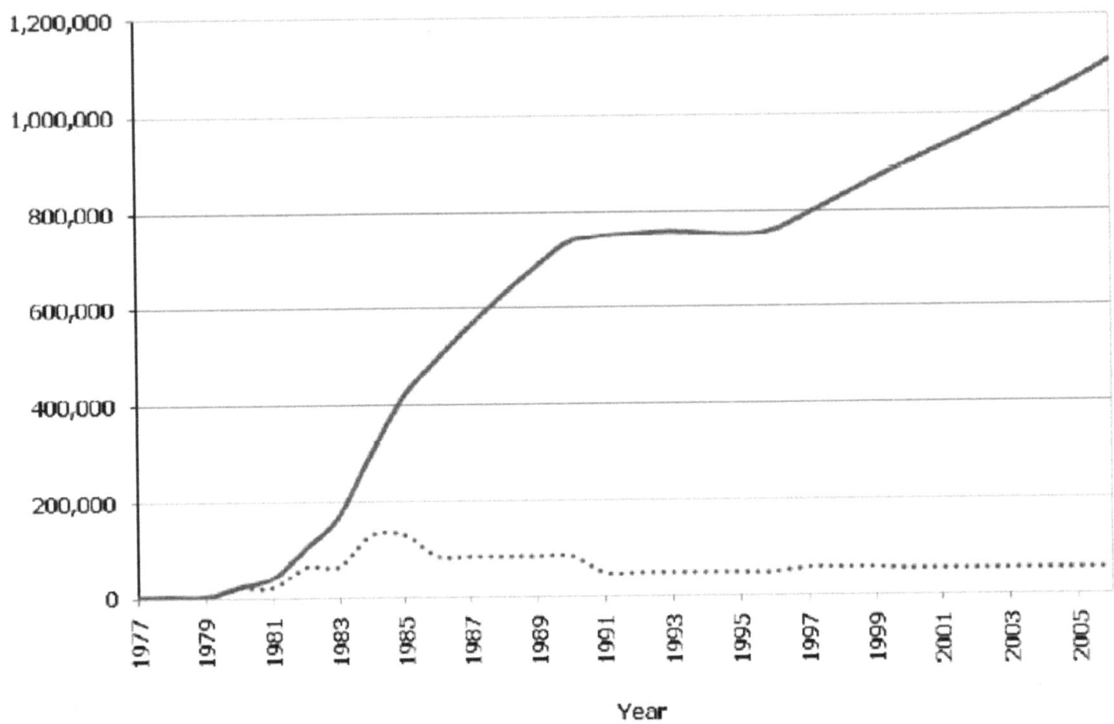

Figure 2-5
Source: *CDC Fact Sheet HIV in the United States, July 2010*

Figure 2-5 shows the prevalence (blue line) and incidence (red dotted line) of HIV. In the early 1980s at the beginning of this epidemic, before we knew what caused AIDS and before we knew that condom use, screening blood donations, and universal precautions by health care personnel could prevent the spread of the virus, the incidence kept going up. More people got infected, and then they in turn infected others. However, we also could not treat HIV initially, and so people would die of AIDS within a few years. The early rise in prevalence is thus attributable solely to the rising incidence. Then we discovered how to prevent new cases: thus, the incidence went down, and while the prevalence took a couple years to catch up, it eventually leveled off. In 1996, access to highly active antiretroviral treatments (HAART) became common,[viii] and it was now possible for people to "live with HIV." The increasing prevalence, starting in the late 1990s, is thus due entirely to an increase in patient survival, or the average duration of illness (you can see that the incidence is steady at that time).

Prevalence therefore comprises 2 characteristics of a disease within a population: the incidence and the average survival time. A change in either one of these components would lead to a change in prevalence; thus, when a change in prevalence is observed, the smart public health professional pauses to consider whether the change is due to a change in the number of new cases (incidence) or to a change in available treatments (and thus survival). One can see that a public health department's response to each of these scenarios would be different.

Summary

This chapter discusses 3 measures of disease frequency: counts, which are used for extremely rare conditions; prevalence, which considers new and existing cases and is used for resource allocation; and incidence, which considers only new cases and is used to study disease **etiology**.

Incidence can further be broken down into *incidence proportion* (which uses the number of people at risk as the denominator) and the *incidence rate* (which uses the sum of the person-time at risk as the denominator). Prevalence is approximately equal to the incidence multiplied by the average survival time after diagnosis.

References

i. Meningococcal disease. Student Health Services. 2009. https://studenthealth.oregonstate.edu/infectious-diseases/meningococcal-disease. Published August 28, 2009. Accessed October 19, 2018. (↲ Return 1) (↲ Return 2)

ii. Oregon Health Authority. Oregon birth data. State of Oregon. https://www.oregon.gov/oha/PH/BirthDeathCertificates/VitalStatistics/birth/Pages/index.aspx. Accessed October 19, 2018.

iii. Oregon Health Authority. Deaths and perinatal deaths data: Annual report volume 2: State of Oregon. https://www.oregon.gov/oha/PH/BIRTHDEATHCERTIFICATES/VITALSTATISTICS/ANNUALREPORTS/VOLUME2/Pages/index.aspx. Accessed October 19, 2018.

iv. Breastfeeding report card, United States, 2013. Centers for Disease Control and Prevention (CDC). 2013. http://www.cdc.gov/breastfeeding/data/reportcard.htm. Accessed September 5, 2014.

v. Percentage of adults aged 65 and over with osteoporosis or low bone mass at the femur neck or lumbar spine: United States, 2005–2010. CDC. https://www.cdc.gov/nchs/data/hestat/osteoporsis/osteoporosis2005_2010.htm. Accessed July 31, 2018. (↲ Return)

vi. American FactFinder—results. Bureau USC. https://factfinder.census.gov/faces/tableservices/jsf/pages/productview.xhtml?pid=PEP_2015_PEPAGESEX&prodType=table. Accessed July 31, 2018. (↲ Return)

vii. Last JM. A *Dictionary of Epidemiology*. 4th ed. New York: Oxford University Press. (↲ Return 1) (↲ Return 2)

viii. Byrne M. A brief history of AZT, HIV's first "ray of hope." Motherboard. 2015. https://motherboard.vice.com/en_us/article/mgb48x/happy-birthday-to-azt-the-first-effective-hiv-treatment. Accessed October 19, 2018. (↲ Return)

3. Surveillance

KELLY JOHNSON AND MARIT L. BOVBJERG

Learning Objectives

After reading this chapter, you will be able to do the following:

1. Define *epidemic* and explain that word's relationship to epidemiology
2. Define *surveillance* and explain how surveillance relates to epidemiology overall
3. Describe some common surveillance systems and methods used in the US
4. Explain the rationale behind notifiable condition reporting, and how this pertains to epidemics, epidemiology, and surveillance

The root word for epidemiology is **epidemic**. An epidemic is "an increase, often sudden, in the number of cases of disease above what is normally expected in that population in that area."[i]

How do we know how much is "expected"? Surveillance!

Public health surveillance is defined by the World Health Organization (WHO) as

> the continuous, systematic collection, analysis, and interpretation of health related data needed for the planning, implementation, and evaluation of public health practice. Such surveillance can (1) serve as an early warning system for impending public health emergencies, (2) document the impact of an intervention, or track progress towards specified goals, and (3) monitor and clarify the epidemiology of health problems, to allow priorities to be set and to inform public health policy and strategies.[ii]

Surveillance activities can be either passive or active. In passive surveillance, the health department *passively* receives reports of suspected injury or illness. Think of this as waiting for disease reports to come to you. Many routine surveillance activities are passive—for instance, systems keeping track of communicable diseases, cancer, and injuries. Epidemiologists collect case reports that are sent to them by health care providers, laboratories, schools, or other entities that are required by law to report this information. In active surveillance, on the other hand, epidemiologists *actively* seek out cases of disease. For example, during an outbreak of salmonellosis associated with a specific source (say, a restaurant), epidemiologists may contact health care providers in the area and ask each for a list of patients seen with symptoms consistent with salmonellosis. These patients are then contacted to see if they were exposed to the suspected source (here, the restaurant). National surveys, such as the National Health and Nutrition

Examination Survey (NHANES),[iii] are also considered active surveillance. The benefit of active surveillance is that it generally results in more complete data, while passive surveillance relies on others (who have numerous duties other than disease reporting) to report cases. The downside to active surveillance is that it is more resource-intensive, with increased personnel and financial requirements.[iv]

Some surveillance activities can be further characterized as population-based. The goal of population-based surveillance is to find every case that occurs within a population, and it is usually part of a mandated effort to collect cases of a specific condition of interest. An example of a condition for which we do population-based surveillance is cancer. Cancer registries aim to capture every case of cancer that occurs in the population the registry covers. This allows clinicians and public health professionals to monitor for trends in diagnoses that might signify a concerning change in the environment and/or for trends in survival that might follow improvements in treatment.

When reporting a specific condition is not required by law, public health officials must estimate incidence and prevalence in other ways. *Sentinel surveillance* involves case reporting from a limited number of hand-picked reporting sites. This type of surveillance is conducted when high-quality data are needed, passive systems are unable to provide these data, and resources are too scarce for complete, population-based active surveillance.[v] For example, annual influenza virus surveillance collects positive influenza specimens from a variety of selected sites each year for **genotyping**. From these specimens, we are able to determine which strains of the influenza virus are circulating each year and thus which strains to include in the annual influenza vaccine. Surveys can also be used to conduct surveillance on a representative sample of the population.

Notifiable Conditions

There is a list of conditions—mostly infectious diseases, but a few chronic diseases and injuries also make the list—that must be reported to the Centers for Disease Control and Prevention (CDC) whenever they are encountered by clinicians or health department officials. For example, say a patient presents to a primary care clinic complaining of high fever, cough, and watery eyes followed by a full-body rash. The nurse practitioner who sees the patient diagnoses measles. This clinic must then report the measles case to the local health department, who in turn reports it to the state health department, who in turn reports it to the CDC. This reporting ideally happens quickly, in a matter of days (or within hours for a potentially major threat).

The list of nationally notifiable conditions is reviewed every year or so and revised according to current public health threats and priorities. For instance, Zika virus and its associated congenital conditions were added to the list in 2016. (The 2020 list of notifiable conditions can be found here.)

Some of the conditions on the list are extremely rare (human rabies, plague) or have even been eradicated (small pox). However, they remain on the notifiable conditions list because in these cases, our expected level (also called the **endemic** level) is 0, and these conditions are dangerous enough that even one suspected case would be cause for an immediate public health intervention.

Each condition on the list has an associated set of case criteria, so after available evidence for a given patient (including laboratory data, symptoms, relevant exposures, and physician diagnoses) is collected, it is compared against the case criteria to either confirm or rule out a given case report. These case criteria are in place to make sure that all epidemiologists are evaluating case reports consistently. For example, the current case criteria for Lyme disease, last revised in 2017, can be found here.

For most of the conditions, the reporting criteria specify that these are new cases so that incidence can be calculated from these data. However, there are exceptions to this—hepatitis C is challenging to identify in its initial stage because few patients exhibit symptoms,[vi] resulting in a large number of hepatitis C infections that are identified during laboratory testing for something unrelated or once symptoms of liver damage occur. For conditions like this, the CDC requests notification of any newly *diagnosed* cases—regardless of whether they are also a new onset case.

The CDC publishes weekly data tabulating all reported cases of the notifiable conditions.[1] Figure 3-1 shows a screenshot of the notifiable conditions tables for meningitis, in Oregon, during the last few months of 2017.[vii]

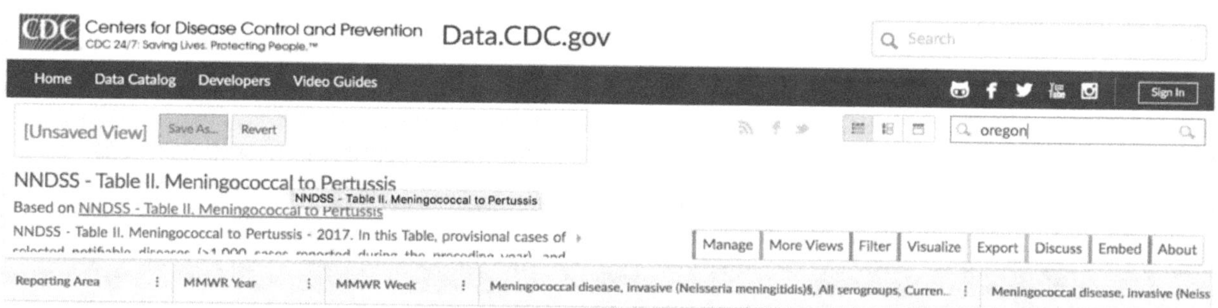

1. It is important to note that underreporting is inherent in all passive surveillance systems, and the notifiable condition data are no different.

OREGON	2017	34	-
OREGON	2017	35	-
OREGON	2017	36	-
OREGON	2017	37	-
OREGON	2017	38	-
OREGON	2017	39	-
OREGON	2017	40	-
OREGON	2017	41	-
OREGON	2017	42	1
OREGON	2017	43	1
OREGON	2017	44	1
OREGON	2017	45	-
OREGON	2017	46	-
OREGON	2017	47	1
OREGON	2017	48	-
OREGON	2017	49	-
OREGON	2017	50	-

Figure 3-1

These cases were part of the epidemic of meningococcal meningitis that occurred at Oregon State University (OSU) during the 2017/2018 academic year—the expected number of cases of meningitis is 0, and the university, after consultation with the local and state health departments, took action (requiring students age 25 and younger to be vaccinated before they could register for classes[viii]) after only 6 cases were reported over the course of several weeks. The complete data tables for the notifiable conditions can be found here.

Notifiable Conditions and Privacy

At this point, most people are familiar with the Health Insurance Portability and Accountability Act (HIPAA) Privacy Rule (45 CFR 154.512[b]). This rule states that your health care provider cannot disclose details about your health or the care that you received (collectively called *protected health information*) without your permission, with some exceptions for insurance and payments, coordinating care with other providers, and so on. Indeed, most clinics require that you acknowledge annually that they have informed you of their HIPAA-compliant privacy policies. Many people are unaware, however, that public health functions such as notifiable condition reporting are exempt from the HIPAA Privacy Rule. Indeed, the US Department of Health and Human Services states on its website, "The HIPAA Privacy Rule recognizes the legitimate need for public health authorities and others responsible for ensuring public health and safety to have access to protected health information to carry out their public health mission...Accordingly, the Rule permits covered entities to disclose protected health information without authorization for specified public health purposes."

Source: https://www.hhs.gov/hipaa/for-professionals/special-topics/public-health/index.html

Cancer Registries

Cancer is a notifiable condition and is worth its own mention, because generally cancer reporting requirements are more extensive than those for other conditions. Depending on the state, a physician who diagnoses a type of cancer (other than non-melanoma skin cancers) must report extensive information to the health department, potentially including the type of tumor, the stage at which it was diagnosed, **histology** information, treatments given, and the eventual outcome (death, recurrence, etc.). Since cancer cases are reported upon diagnosis,[2] this can also be a potential source of incidence data, with the same caveats discussed above for hepatitis C (i.e., that sometimes a diagnosis occurs quite late in the disease process). Cancer registries are somewhat unique compared to other notifiable conditions data because patients are followed over time. Several states contribute their cancer registry data to the Surveillance, Epidemiology, and End Results (SEER) database,[ix] which is available for both surveillance and research purposes.

Vital Statistics

Birth and death certificates—together called *vital statistics*[x] —constitute another ongoing surveillance system. Local hospitals report births and deaths up to their state health departments, who in turn report to the CDC. By keeping track of the health of newborns and childbearing women, as well as causes of death for everyone, public health officials can spot potential emerging trends that would warrant intervention. Annual reports summarizing all births and deaths that occurred in the US, as well as any notable changes from previous years, can be found on the CDC's website here.

2. Cancer registries also include cases identified through autopsies/death certificates that were not diagnosed while the person was alive. Since not everyone has an autopsy when they die, undiagnosed cancers among people who have died are underreported.

Survey-Based Surveillance Systems

The US conducts numerous surveillance activities that involve direct data collection from individual residents, usually via questionnaires, although NHANES includes physical exam and laboratory data as well. Other examples of surveillance include the Behavioral Risk Factor Surveillance System (BRFSS),[xi] which is a telephone-based survey of adults, who are asked to self-report their health and health behaviors; and the Pregnancy Risk Assessment Monitoring System (PRAMS),[xii] which is a paper-based survey of women who have recently given birth, who report on health and health care utilization for themselves and their newborn(s). Data from these surveys are used by public health professionals to monitor trends over time—for instance, the map on seat belt use shown in chapter 1 was made using BRFSS data—but they contain only data from prevalent cases. On the plus side, the survey data are freely available to students and researchers, and numerous articles are published each year using these datasets.

Conclusions

Surveillance activities allow epidemiologists and other public health professionals to monitor the "usual" levels of disease in a population. The ultimate goal of surveillance is to notice potential public health threats early so that a proper response can be mounted before a public health crisis ensues. The US has numerous surveillance systems operating at any one time, and much of the benefit from these systems develops as data are compared over time.

References

i. Principles of epidemiology: Lesson 1—section 11. Centers for Disease Control and Prevention (CDC). https://www.cdc.gov/ophss/csels/dsepd/ss1978/lesson1/section11.html. Accessed September 10, 2018. (↲ Return)

ii. Public health surveillance. World Health Organization (WHO). http://www.who.int/topics/public_health_surveillance/en/. Accessed September 10, 2018. (↲ Return)

iii. National Health and Nutrition Examination Survey Homepage. CDC. 2018. https://www.cdc.gov/nchs/nhanes/index.htm. Accessed October 19, 2018. (↲ Return)

iv. Groseclose S, Sullivan K, Gibbs N, Knowles C. Management of the surveillance information system and quality control data. In: Teutsch S, Churchill R, eds. *Principles and Practice of Public Health Surveillance*. Vol 2. New York: Oxford University Press; 2000:95-111. (↲ Return)

v. Sentinel surveillance. WHO. http://www.who.int/immunization/monitoring_surveillance/burden/vpd/surveillance_type/sentinel/en/. Accessed September 21, 2018. (↲ Return)

vi. Hepatitis C questions and answers for the public. CDC. 2018. https://www.cdc.gov/hepatitis/hcv/cfaq.htm. Accessed September 21, 2018. (↲ Return)

vii. Calgary O. Meningococcal to pertussis: NNDSS—table II. CDC. https://data.cdc.gov/NNDSS/NNDSS-Table-II-Meningococcal-to-Pertussis/hatw-7gqy/data. Accessed September 10, 2018. (↲ Return)

viii. Meningococcal disease. Student Health Services. 2009. https://studenthealth.oregonstate.edu/infectious-diseases/meningococcal-disease. Accessed October 19, 2018. (↲ Return)

ix. Surveillance, epidemiology, and end results program. National Cancer Institute. https://seer.cancer.gov/. Accessed October 20, 2018. (↲ Return)

x. National Vital Statistics System homepage: NVSS. CDC. 2018. https://www.cdc.gov/nchs/nvss/index.htm. Accessed October 20, 2018. (↲ Return)

xi. BRFSS—Behavioral Risk Factor Surveillance System. CDC. http://www.cdc.gov/brfss/. Accessed March 3, 2015. (↲ Return)

xii. Pregnancy Risk Assessment Monitoring System—reproductive health. CDC. http://www.cdc.gov/prams/. Accessed September 5, 2014. (↲ Return)

4. Introduction to 2 x 2 Tables, Epidemiologic Study Design, and Measures of Association

Learning Objectives

After reading this chapter, you will be able to do the following:

1. Interpret data found in a 2 x 2 table
2. Compare and contrast the 4 most common types of epidemiologic studies: cohort studies, randomized controlled trials, case-control studies, and cross-sectional studies
3. Calculate and interpret relative measures of association (risk ratios, rate ratios, odds ratios)
4. Explain which measures are preferred for which study designs and why
5. Discuss the differences between absolute and relative measures of association

In epidemiology, we are often concerned with the degree to which a particular exposure might cause (or prevent) a particular disease. As detailed later in chapter 10, it is difficult to claim causal effects from a single epidemiologic study; therefore, we say instead that exposures and diseases are (or are not) statistically *associated*. This means that the exposure is **disproportionately distributed** between individuals with and without the disease. The degree to which exposures and health outcomes are associated is conveyed through a **measure of association**. Which measure of association to choose depends on whether you are working with **incidence** or **prevalence** data, which in turn depends on the type of study design used. This chapter will therefore provide a brief outline of common epidemiologic study designs interwoven with a discussion of the appropriate measure(s) of association for each. In chapter 9, we will return to study designs for a more in-depth discussion of their strengths and weaknesses.

Necessary First Step: 2 x 2 Notation

Before getting into study designs and measures of association, it is important to understand the notation used in epidemiology to convey exposure and disease data: the **2 x 2 table**. A 2 x 2 table (or *two-by-two table*) is a compact summary of data for 2 variables from a study—namely, the exposure and the health outcome. Say we do a 10-person study on smoking and **hypertension**, and collect the following data, where Y indicates yes and N indicates no:

Table 4-1

Participant #	Smoker?	Hypertension?
1	Y	Y
2	Y	N
3	Y	Y
4	Y	Y
5	N	N
6	N	Y
7	N	N
8	N	N
9	N	Y
10	N	N

You can see that we have 4 smokers, 6 nonsmokers, 5 individuals with hypertension, and 5 without. In this example, smoking is the exposure and hypertension is the health outcome, so we say that the 4 smokers are "exposed" (E+), the 6 nonsmokers are "unexposed" (E−), the 5 people with hypertension are "diseased" (D+), and the 5 people without hypertension are "nondiseased" (D−). This information can be organized into a 2 × 2 table:

Table 4-2

	D+	D-
E+	3	1
E-	2	4

The 2 × 2 table summarizes the information from the longer table above so that you can quickly see that 3 individuals were both exposed and diseased (persons 1, 3, and 4); one individual was exposed but not diseased (person 2); two individuals were unexposed but diseased (persons 6 and 9); and the remaining 4 individuals were neither exposed nor diseased (persons 5, 7, 8, and 10). Though it does not really matter whether exposure or disease is placed on the left or across the top of a 2 × 2 table, the convention in epidemiology is to have exposure on the left and disease across the top.

When discussing 2 x 2 tables, epidemiologists use the following shorthand to refer to specific cells:

Table 4-3

	D+	D-
E+	A	B
E-	C	D

It is often helpful to calculate the margin totals for a 2 x 2 table:

Table 4-4

	D+	D-	Total
E+	3	1	4
E-	2	4	6
Total	5	5	10

Or:

Table 4-5

	D+	D-	Total
E+	A	B	A+B
E-	C	D	C+D
Total	A+C	B+D	A+B+C+D

The margin totals are sometimes helpful when calculating various measures of association (and to check yourself against the original data).

Continuous versus Categorical Variables

Continuous variables are things such as age or height, where the possible values for a given person are infinite, or close to it. Categorical variables are things such as religion or favorite color, where there is a discrete list of possible answers. Dichotomous variables are a special case of categorical variable where there are only 2 possible answers. It is possible to dichotomize a continuous variable—if you have an "age" variable, you could split it into "old" and "young." However, is it not always advisable to do this because a lot of information is lost.

Furthermore, how does one decide where to dichotomize? Does "old" start at 40, or 65? Epidemiologists usually prefer to leave continuous variables continuous to avoid having to make these judgment calls.

Nonetheless, having dichotomous variables (a person is either exposed or not, either diseased or not) makes the math much easier to understand. For the purposes of this book, then, we will assume that all exposure and disease data can be meaningfully dichotomized and placed into 2×2 tables.

Studies That Use Incidence Data

Cohorts

There are 4 types of epidemiologic studies that will be covered in this book,[1] two of which collect incidence data: **prospective cohort studies** and **randomized controlled trials.** Since these study designs use incidence data, we instantly know 3 things about these study types. One, we are looking for new cases of disease. Two, there is thus some longitudinal follow-up that must occur to allow for these new cases to develop. Three, we must start with those who were at risk (i.e., without the disease or health outcome) as our **baseline**.

The procedure for a prospective cohort study (hereafter referred to as just a "**cohort study**," though see the inset box on **retrospective cohort studies** later in this chapter) begins with the **target population**, which contains both diseased and non-diseased individuals:

1. These 4 study designs are the basis for nearly all others (e.g., case-crossover studies are a subtype of case-control studies). A few additional designs are covered in chapter 9, but a firm understanding of the 4 designs covered in this chapter will set beginning epidemiology students up to be able to critically read essentially all of the literature.

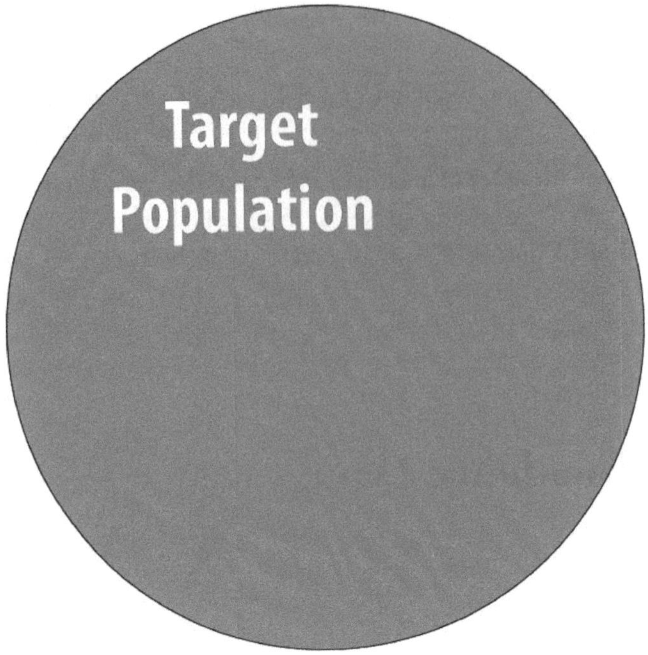

Figure 4-1

As discussed in chapter 1, we rarely conduct studies on entire populations because they are too big for it to be logistically feasible to study everyone in the population. Therefore we draw a **sample** and perform the study with the individuals in the sample. For a cohort study, since we will be calculating incidence, we must start with individuals who are at risk of the outcome. We thus draw a non-diseased sample from the target population:

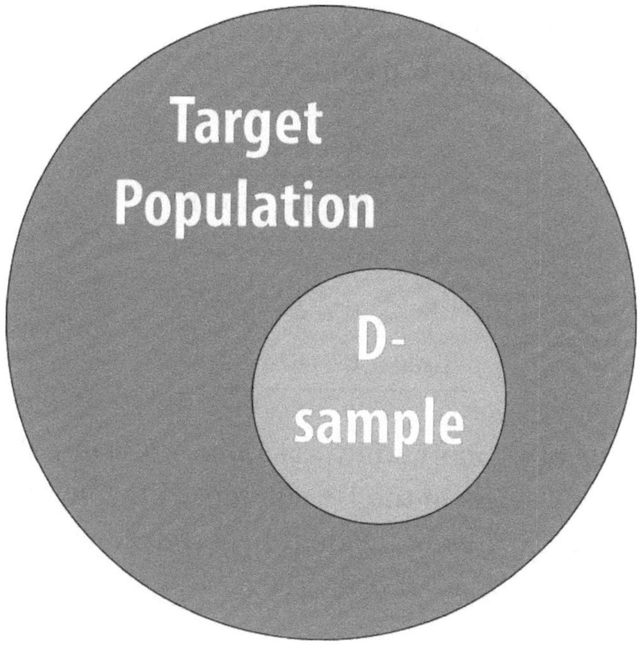

Figure 4-2

The next step is to assess the exposure status of the individuals in our sample and determine whether they are exposed or not:

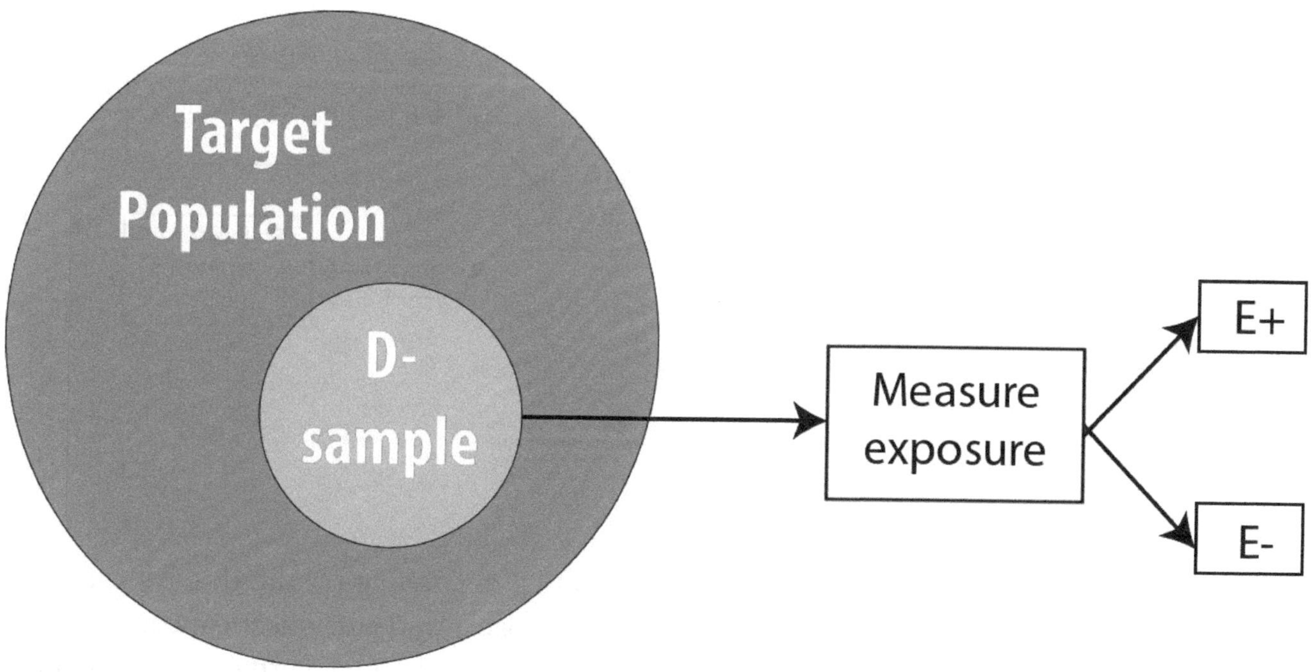

Figure 4-3

After assessing which participants were exposed, our 2 x 2 table (using the 10-person smoking/HTN data example from above) would look like this:

Table 4-6

	D+	D-	Total
E+	0	4	4
E-	0	6	6
Total	0	10	10

By definition, at the beginning of a cohort study, *everyone is still at risk* of developing the disease, and therefore there are no individuals in the D+ column. In this hypothetical example, based on the data above, we will observe 5 cases of incident hypertension as the study progresses–but at the beginning, none of these cases have yet occurred.

We then follow the participants in our study for some length of time and observe incident cases as they arise.

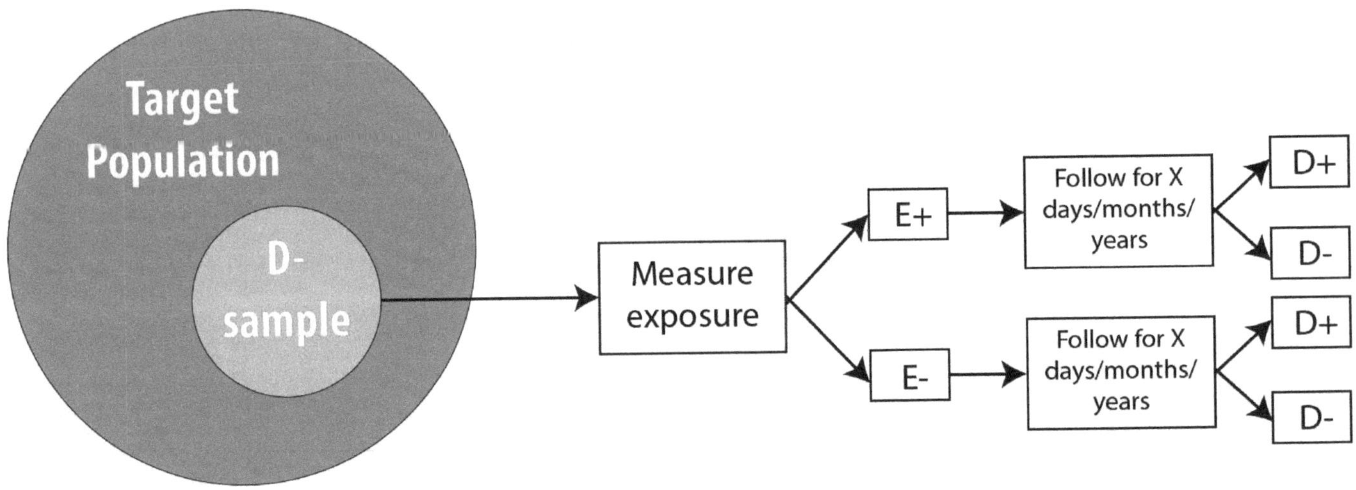

Figure 4-4

As mentioned in chapter 2, the length of follow-up varies depending on the disease process in question. For a research question regarding childhood exposure and late-onset cancer, the length of follow-up would be decades. For an infectious disease outbreak, the length of follow-up might be a matter of days or even hours, depending on the **incubation period** of the particular disease.

Assuming we are calculating **incidence proportions** (which use the number of people at risk in the denominator) in our cohort, our 2 × 2 table at the end of the smoking/HTN study would look like this:

Table 4-7

	D+	D-	Total
E+	3	1	4
E-	2	4	6
Total	5	5	10

It is important to recognize that when epidemiologists talk about a 2 × 2 table from a cohort study, they mean the 2 × 2 table at the *end* of the study—the 2 × 2 table from the beginning was much less interesting, as the D+ column was empty!

From this 2 × 2 table, we can calculate a number of useful measures, detailed below.

Calculating the Risk Ratio from the Hypothetical Smoking/Hypertension Cohort Study

We can start by calculating the overall incidence of disease in our sample (assume that our smoking/HTN study included 10 years of follow-up):

Incidence proportion = $\frac{\text{number of new cases}}{\text{population at risk at baseline}} = \frac{5}{10}$ = 50 cases per 100 people in 10 years

Using ABCD notation for a 2 x 2 table, the formula for the overall incidence proportion is:

$$\frac{(A+C)}{(A+B+C+D)}$$

We can also calculate the incidence only among exposed individuals:

$I_{E+} = \frac{A}{(A+B)} = \frac{3}{4}$ = 75 per 100 in 10 years

Likewise, we can calculate the incidence only among unexposed individuals:

$I_{E-} = \frac{C}{(C+D)} = \frac{2}{6}$ = 33 per 100 in 10 years

Recall that our original goal with the cohort study was to see whether exposure is associated with disease. We thus need to compare the I_{E+} to the I_{E-}. The most common way of doing this is to calculate their combined ratio:

Risk Ratio = $\frac{I_{E+}}{I_{E-}} = \frac{\text{75 per 100 in 10 years}}{\text{33 per 100 in 10 years}}$ = 2.27

Using ABCD notation, the formula for RR is:

$$\frac{\frac{A}{(A+B)}}{\frac{C}{(C+D)}}$$

Note that risk ratios (RR) have no units, because the time-dependent units for the 2 incidences cancel out.

If the RR is greater than 1, it means that we observed more disease in the exposed group than in the unexposed group. Likewise, if the RR is less than 1, it means that we observed less disease in the exposed group than in the unexposed group. If we assume causality, an exposure with an RR < 1 is *preventing* disease, and an exposure with an RR > 1 is *causing* disease. The **null value** for a risk ratio is 1.0, which would mean that there was no observed association between exposure and disease. You can see how this would be the case—if the incidence was identical in the exposed and unexposed groups, then the RR would be 1, since x divided by x is 1.

Because the null value is 1.0, one must be careful if using the words *higher* or *lower* when interpreting RRs. For instance, an RR of 2.0 means that the disease is twice as common, or twice as high, in the exposed compared to the unexposed—not that it is 2 times *more* common, or 2 times *higher*, which would be an RR of 3.0 (since the null value is 1, not 0). If you do not see the distinction between these, don't sweat it—just memorize and use the template sentence below, and your interpretation will be correct.

The correct interpretation of an RR is:

"The risk of [disease] was [RR] *times as high* in [exposed] compared to [unexposed] over [x] days/months/years."

Using our smoking/HTN example:

"The risk of hypertension was 2.27 times as high in smokers compared to nonsmokers over 10 years."

The key phrase is *times as high*; with it, the template sentence works regardless of whether the RR is above or below 1. For an RR of 0.5, saying "0.5 times as high" means that you multiply the risk in the unexposed by 0.5 to get the risk in the exposed, yielding a *lower* incidence in the exposed—as one expects with an RR < 1.

If our cohort study instead used a person-time approach, the 2 x 2 table at the end of the study would have a column for sum of the **person-time at risk (PTAR)**:

Table 4-8

	D+	D-	Total	Σ PTAR
E+	3	1	4	27.3 PY
E-	2	4	6	52.9 PY
Total	5	5	10	80.2 PY

Calculating the Rate Ratio from the Hypothetical Smoking/Hypertension Cohort Study

Using a person-time denominator, the **incidence rate** for the overall study is:

$$I = \frac{\textbf{5 new cases}}{\textbf{80.2 PY}} = 6.2 \text{ per 100 person-years}$$

Likewise, the incidence rate among exposed persons is:

$$I_{E+} = \frac{\textbf{3}}{\textbf{27.3}} = 11.0 \text{ per 100 person-years}$$

And the incidence among unexposed persons is:

$$I_{E+} = \frac{\textbf{2}}{\textbf{59.2}} = 3.8 \text{ per 100 person-years}$$

We again take the ratio of incidence in the exposed to incidence in the unexposed, this time calculating a **rate ratio** (also abbreviated RR):

$$RR = \frac{I_{E+}}{I_{E-}} = 2.9$$

As when using incidence proportions, the units cancel out, and we are left with just a number.

The interpretation is the same as it would be for the risk ratio; one just needs to substitute the word *rate* for the word *risk:*

> The rate of hypertension was 2.9 times as high in smokers compared to non-smokers, over 10 years.

Notice that the interpretation sentence still includes the duration of the study, even though some individuals (the 4 who developed hypertension) were **censored** before that time. This is because knowing how long people were followed for (and thus given time to develop disease) is still important when interpreting the findings. As discussed in chapter 2, 100 years of person-time can be accumulated in any number of different ways; knowing that the duration of the study was 10 years (rather than 1 year or 50 years) might make a difference in terms of how (or if) one applies the findings in practice.

"Relative Risk"

Both the risk ratio and the rate ratio are abbreviated RR. This abbreviation (and the risk ratio and/or rate ratio) is often referred to by epidemiologists as *relative risk*. This is an example of inconsistent lexicon in the field of epidemiology; in this book, I use *risk ratio* and *rate ratio* separately (rather than relative risk as an umbrella term) because it is helpful, in my opinion, to distinguish between studies using the population at risk vs. those using a person-time at risk approach. Regardless, a measure of association called RR is always calculated as incidence in the exposed divided by incidence in the unexposed.

Retrospective Cohort Studies

Throughout this book, I will focus on prospective cohort studies. One can also conduct a *retrospective* cohort study, mentioned here because public health and clinical practitioners will encounter retrospective cohort studies in the literature. In theory, a retrospective cohort study is conducted exactly like a prospective cohort study: one begins with a non-diseased sample from the target population, determines who was exposed, and "follows" the sample for *x* days/months/years, looking for incident cases of disease. The difference is that, for a retrospective cohort study, all this has already happened, and one reconstructs this information using existing records. The most common way to do retrospective cohort studies is by using employment records (which often have job descriptions useful for surmising exposure—for instance, the floor manager was probably exposed to whatever chemicals were on the factory floor, whereas human resource officers probably were not), medical records, or other administrative datasets (e.g., military records).

Continuing with our smoking/HTN 10-year cohort example, one might do a retrospective cohort using medical records as follows:

- Go back to all the records from 10 years ago and determine who already had hypertension (these people are not at risk and are therefore not eligible) or otherwise does not meet the sample inclusion criteria
- Determine, among those at risk 10 years ago, which individuals were smokers
- Determine which members of the sample then developed hypertension during the intervening 10 years

Retrospective cohorts are analyzed just like prospective cohorts—that is, by calculating rate ratios or risk ratios. However, for beginning epidemiology students, retrospective cohorts are often confused with case-control studies; therefore we will focus exclusively on prospective cohorts for the remainder of this book. (Indeed, occasionally even seasoned scientists are confused about the difference!)[i]

Randomized Controlled Trials

The procedure for a randomized controlled trial (RCT) is exactly the same as the procedure for a prospective cohort, with one exception: instead of allowing participants to self-select into "exposed" and "unexposed" groups, the investigator in an RCT randomly assigns some participants (usually half) to "exposed" and the other half to "unexposed." In other words, exposure status is determined entirely by chance. This is the type of study required by the Food and Drug Administration for approval of new drugs: half of the participants in the study are randomly assigned to the new drug and half to the old drug (or to a placebo, if the drug is intended to treat something previously untreatable). The diagram for an RCT is as follows:

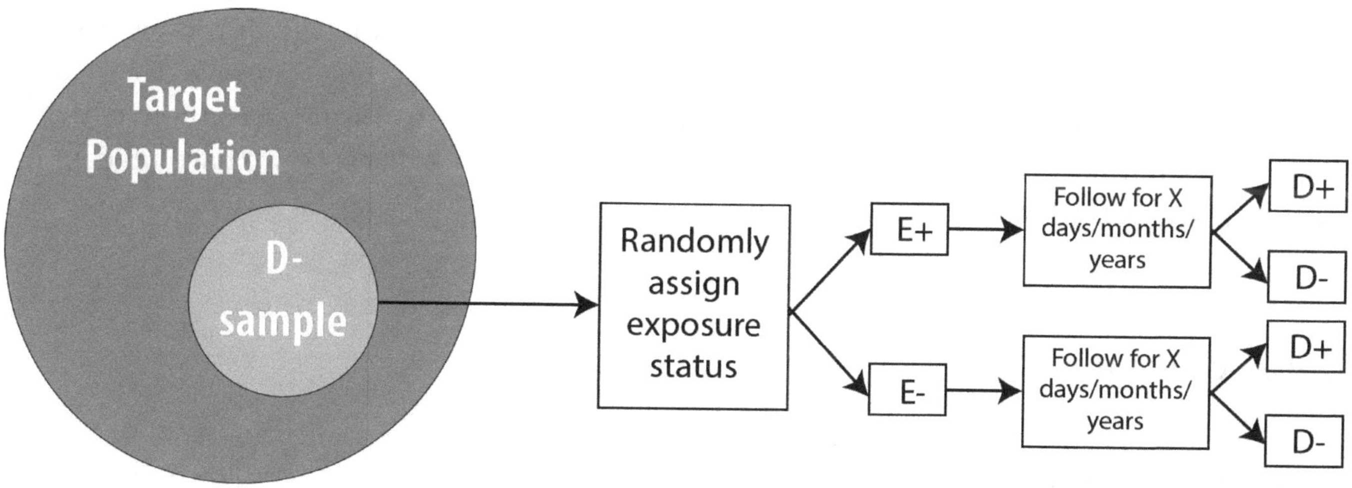

Figure 4-5

Note that the *only difference* between an RCT and a prospective cohort is the first box: instead of measuring existing exposures, we now tell people whether they will be exposed or not. We are still measuring incident disease, and we are therefore still calculating either the risk ratio or the rate ratio.

Observational versus Experimental Studies

Cohort studies are a subclass of **observational studies**, meaning the researcher is merely observing what happens in real life—people in the study self-select into being exposed or not depending on their personal preferences and life circumstances. The researcher then measures and records a given person's level of exposure. Cross-sectional and case-control studies are also observational. Randomized controlled trials, on the other hand, are *experimental* studies—the researcher is conducting an experiment that involves telling people whether they will be exposed to a condition or not (e.g., to a new drug).

Studies That Use Prevalence Data

Following participants while waiting for incident cases of disease is expensive and time-consuming. Often, epidemiologists need a faster (and cheaper) answer to their question about a particular exposure/disease combination. One might instead take advantage of prevalent cases of disease, which by definition have already occurred and therefore require no wait. There are 2 such designs that I will cover: **cross-sectional studies** and **case-control studies**. For both of these, since we are not using incident cases, we cannot calculate the RR, because we have no data on incidence. We instead calculate the **odds ratio (OR)**.

Cross-sectional

Cross-sectional studies are often referred to as *snapshot* or *prevalence* studies: one takes a "snapshot" at a particular point in time, determining who is exposed and who is diseased simultaneously. The following is a visual:

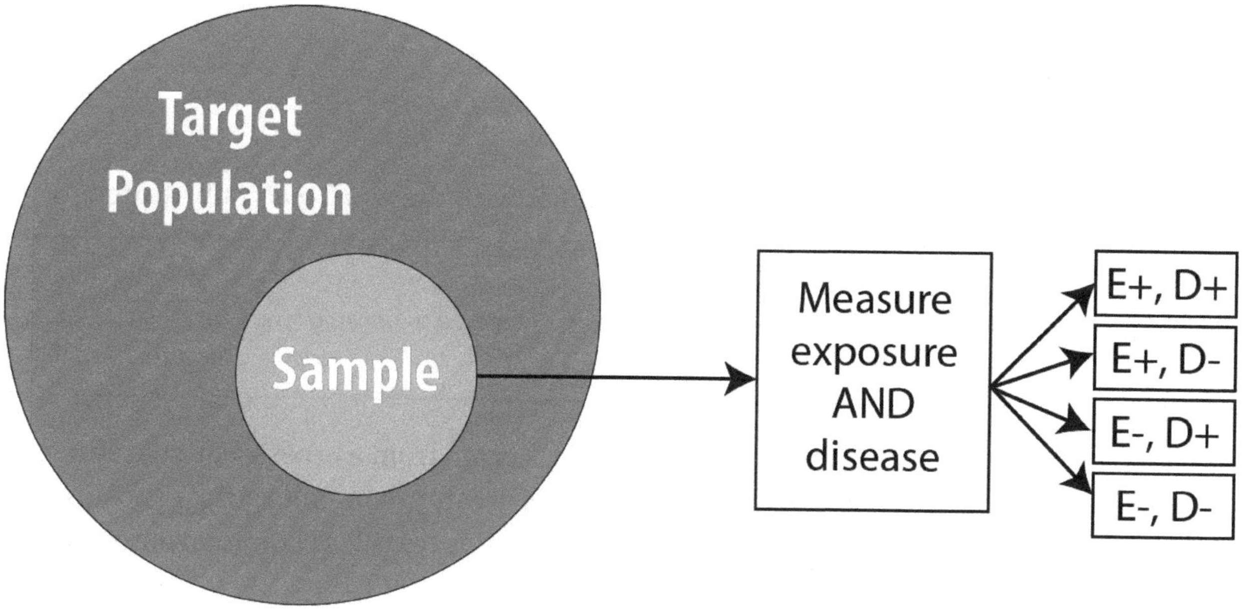

Figure 4-6

Note that the sample is now no longer composed entirely of those at risk because we are using prevalent cases—thus by definition, some proportion of the sample will be diseased **at baseline**. As mentioned, we cannot calculate the RR in this scenario, so instead we calculate the OR.

Calculating the Odds Ratio from the Hypothetical Smoking/Hypertension Cross-Sectional Study

The formula for OR for a cross-sectional study is:

$$\text{OR} = \frac{\textbf{odds of disease in the exposed group}}{\textbf{odds of disease in the unexposed group}}$$

The *odds* of an event is defined statistically as the number of people who experienced an event divided by the number of people who did not experience it. Using 2 × 2 notation, the formula for OR is:

$$\text{OR} = \frac{\frac{A}{B}}{\frac{C}{D}} = \frac{AD}{BC}$$

For our smoking/HTN example, if we assume those data came from a cross-sectional study, the OR would be:

$$\text{OR} = \frac{\frac{3}{1}}{\frac{2}{4}} = \frac{3 * 4}{2 * 1} = 6.0$$

Again there are no units.

The interpretation of an OR is the same as that of an RR, with the word *odds* substituted for *risk*:

The odds of hypertension were 6.0 times as high in smokers compared to nonsmokers.

Note that we now no longer mention time, as these data came from a cross-sectional study, which does not involve time. As with interpretation of RRs, ORs greater than 1 mean the exposure is more common among diseased, and ORs less than 1 mean the exposure is less common among diseased. The null value is again 1.0.

For 2 x 2 tables from cross-sectional studies, one can additionally calculate the overall *prevalence* of disease as

$$\text{Prevalence} = \frac{(A+C)}{(A+B+C+D)}$$

Finally, some authors will refer to the OR in a cross-sectional study as the *prevalence odds ratio*—presumably, just as a reminder that cross-sectional studies are conducted on prevalent cases. The calculation of such a measure is exactly the same as the OR as presented above.

OR *versus* RR

As you can see from the (hypothetical) example data in this chapter, the OR will always be further from the null value than the RR. The more common the disease, the more this is true. If the disease has a prevalence of about 5% or less, then the OR does provide a close approximation of the RR; however, as the disease in question becomes more common (as in this example, with a hypertension prevalence of 40%), the OR deviates further and further from the RR.

Occasionally, you will see a cohort study (or very rarely, an RCT) that reports the OR instead of the RR. Technically this is not correct, because cohorts and RCTs use incident cases, so the best choice for a measure of association is the RR. However, one common statistical modeling technique—logistic regression—automatically calculates ORs. While it is possible to back-calculate the RR from these numbers, often investigators do not bother and instead just report the OR. This is troublesome for a couple of reasons: first, it is easier for human brains to interpret risks as opposed to odds, and therefore risks should be used when possible; and second, cohort studies and RCTs almost always have relatively common outcomes (see chapter 9), thus reporting the OR makes it seem as if the exposure is a bigger problem (or a better solution, if OR < 1) than it "really" is.

Case-Control

The final type of epidemiologic study that is commonly used is the case-control study. It also begins with prevalent cases and thus is faster and cheaper than longitudinal (prospective cohort or RCT) designs. To conduct a case-control study, one first draws a sample of diseased individuals (cases):

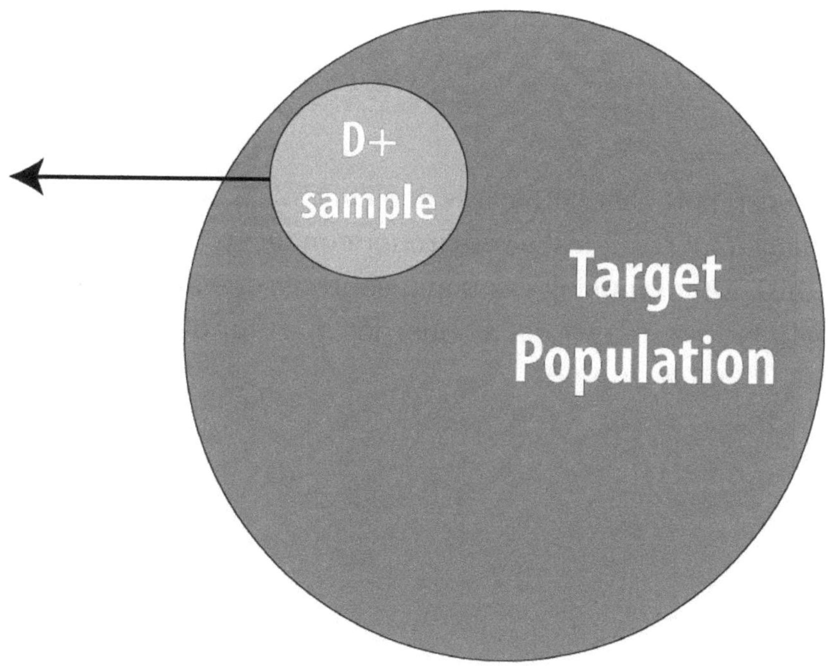

Figure 4-7

Then a sample of nondiseased individuals (controls):

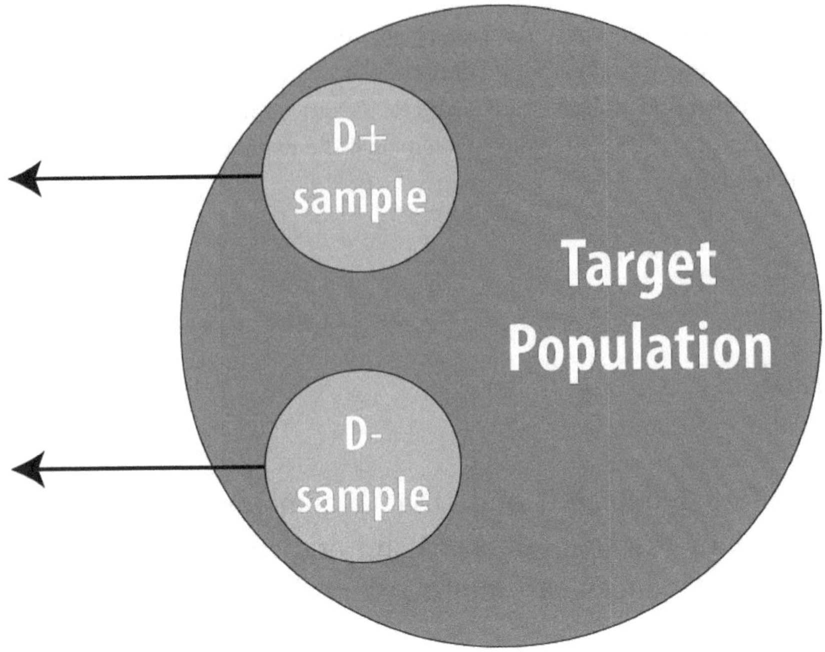

Figure 4-8

First and foremost, note that both cases and controls come from the same underlying population. This is extremely important, lest a researcher conduct a biased case-control study (see chapter 9 for more on this). After sampling cases and controls, one measures exposures at *some point in the past*. This might be yesterday (for a foodborne illness) or decades ago (for osteoporosis):

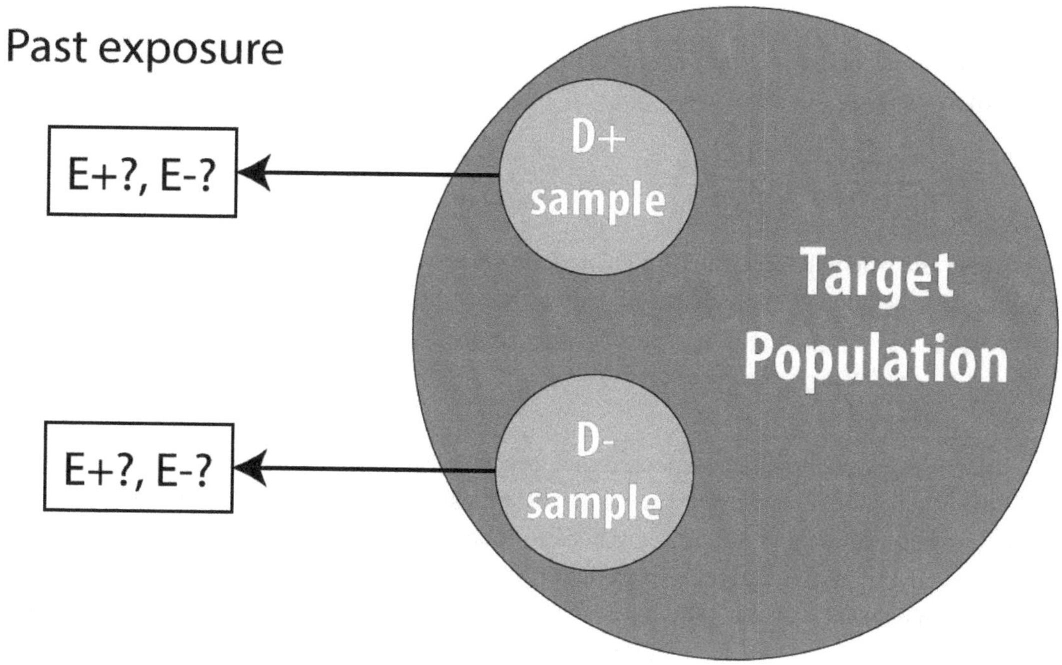

Figure 4-9

Again, we cannot calculate incidence because we are using prevalent cases, so instead we calculate the OR in the same manner as above. The interpretation is identical, but now we must refer to the time period because we explicitly looked at past exposure data:

> The odds of hypertension are 6.0 times as high in people who were smokers 10 years ago, compared to people who were nonsmokers 10 years ago.

Note, however, that one cannot calculate the overall sample prevalence using a 2 × 2 table from a case-control study, because we artificially set the prevalence in our sample (usually at 50%) by deliberately choosing individuals who were diseased for our cases.

Exposure OR versus Disease OR

Technically, for a case-control study, one calculates the *disease* OR rather than the *exposure* OR (which is presented under cross-sectional studies). In other words, since in case-control studies we begin with disease, we are calculating the odds of being exposed among those who are diseased compared to the odds of being exposed among those who are not diseased:

$$OR_{disease} = \frac{(A/C)}{(B/D)} = \frac{AD}{BC}$$

The exposure odds ratio, you will remember, calculates the odds of being diseased among those who are exposed, compared to the odds of being diseased among those who are unexposed:

$$OR_{exposure} = \frac{(A/B)}{(C/D)} = \frac{AD}{BC}$$

In advanced epidemiology classes, one is expected to appreciate the nuances of this difference and to articulate the rationale behind it. However, since both the exposure and the disease odds ratios simplify to the same final equation, here we will not differentiate between them. The interpretation is the same: an OR > 1 means that disease is more common in the exposed group (or exposure is more common in the diseased group—same thing), and an OR < 1 means that disease is less common in the exposed group (or exposure is less common in the diseased group—again, same thing).

Risk Difference

RR and OR are known as *relative* or *ratio* measures of association for obvious reasons. These measures can be misleading, however, if the **absolute risks** (incidences) are small.[2] For example, if a cohort study was done, and investigators observed an incidence in the exposed of 1 per 1,000,000 in 20 years and an incidence in the unexposed, and an incidence in the unexposed of 2 per 1,000,000 in 20 years, the RR would be 0.5: there is a 50% reduction in disease in the exposed group. Break out the public health intervention! However, this *ratio* measure masks an important truth: the *absolute* difference in risk is quite small: 1 in a million.

To address this issue, epidemiologists sometimes calculate instead the risk difference instead:

$$RD = I_{E+} - I_{E-}$$

Unfortunately, this **absolute measure of association** is not often seen in the literature, perhaps because interpretation implies causation more explicitly or because it is more difficult to control for confounding variables (see chapter 7) when calculating difference measures.

Regardless, in our smoking/HTN example, the RD is:

$$RD = I_{E+} - I_{E-} = 75 \text{ per 100 in 10 years} - 33 \text{ per 100 in 10 years} = 42 \text{ per 100 in 10 years}$$

2. Declercq E. The absolute power of relative risk in debates on repeat cesareans and home birth in the United States. *J Clin Ethics*. 2013;24(3):215-224

Note that the RD has the same units as incidence, since units do not cancel when subtracting. The interpretation is as follows:

> Over 10 years, the excess number of cases of HTN attributable to smoking is 42; the remaining 33 would have occurred anyway.

You can see how this interpretation assigns a more explicitly causal role to the exposure.

More common (but still not nearly as common as the ratio measures) are a pair of measures derived from the RD: the attributable risk (AR) and the number needed to treat/number needed to harm (NNT/NNH).

The AR is calculated as RD/I_{E+}. Here,

AR = 42 per 100 in 10 years / 75 per 100 in 10 years = 56%

Interpretation:

> 56% of cases can be attributed to smoking, and the rest would have happened anyway.

Again this implies causality; furthermore, because diseases all have more than one cause (see chapter 10), the ARs for each possible cause will sum to well over 100%, making this measure less useful.

Finally, calculating NNT/NNH (both of which are similar, with the former being for preventive exposures and the latter for harmful ones) is simple:

$$NNT = 1/RD$$

In our example,

NNH = 1 / 42 per 100 per 10 years = 1/0.42 per 10 years = 2.4

Interpretation:

> Over 10 years, for every 2.4 smokers, 1 will develop hypertension.

For a protective exposure, the NNT (commonly used in clinical circles) is interpreted as the number you need to treat in order to prevent one case of a bad outcome. For harmful exposures, as in our smoking/HTN example, it is the number needed to be exposed to cause one bad outcome. For many drugs in common use, the NNTs are in the hundreds or even thousands.[iii][iv]

Conclusions

Epidemiologic data are often summarized in 2 × 2 tables. There are 2 main measures of association commonly used in epidemiology: the risk ratio/rate ratio (relative risk) and the odds ratio. The former is calculated for study designs that collect data on incidence: cohorts and RCTs. The latter is calculated for study designs that use prevalent cases: cross-sectional studies and case-control studies. Absolute measures of association (e.g., risk difference) are not seen as often in epidemiologic literature, but it is nonetheless always important to keep the absolute risks (incidences) in mind when interpreting results.

Below is a table summarizing the concepts from this chapter:

Study Design	Methods Summary	Incident or Prevalent Cases?	Preferred Measure of Association
Cohort	Start with a nondiseased sample, determine exposure, follow over time.	Incident	Risk ratio or rate ratio
RCT	Start with a nondiseased sample, assign exposure, follow over time	Incident	Risk ration or rate ratio
Case-Control	Start with diseased (cases), recruit comparable nondiseased (controls), look at previous exposures	Prevalent	Odds ratio
Cross-sectional	From a sample, assess both exposure status and disease status simultaneously	Prevalent	Odds ratio

---

References

i. Bodner K, Bodner-Adler B, Wierrani F, Mayerhofer K, Fousek C, Niedermayr A, Grünberger. Effects of water birth on maternal and neonatal outcomes. *Wien Klin Wochenschr.* 2002;114(10-11):391-395. (↲ Return)

ii. Declercq E. The absolute power of relative risk in debates on repeat cesareans and home birth in the United States. *J Clin Ethics.* 2013;24(3):215-224.

iii. Mørch LS, Skovlund CW, Hannaford PC, Iversen L, Fielding S, Lidegaard Ø. Contemporary hormonal contraception and the risk of breast cancer. N *Engl J Med.* 2017;377(23):2228-2239. doi:10.1056/NEJMoa1700732 (↲ Return)

iv. Brisson M, Van de Velde N, De Wals P, Boily M-C. Estimating the number needed to vaccinate to prevent diseases and death related to human papillomavirus infection. CMAJ *Can Med Assoc J.* 2007;177(5):464-468. doi:10.1503/cmaj.061709 (↲ Return)

5. Random Error

Learning Objectives

After reading this chapter, you will be able to do the following:

1. Define *random error* and differentiate it from bias
2. Illustrate random error with examples
3. Interpret a *p*-value
4. Interpret a confidence interval
5. Differentiate between type 1 and type 2 statistical errors and explain how they apply to epidemiologic research
6. Describe how statistical power affects research

In this chapter, we will cover **random error**—where it comes from, how we deal with it, and what it means for epidemiology.

What Is Random Error?

First and foremost, random error is not **bias**. Bias is systematic error and is covered in further detail in chapter 6.

Random error is just what it sounds like: random errors in the data. All data contain random errors, because no measurement system is perfect. The magnitude of random errors depends partly on the scale on which something is measured (errors in molecular-level measurements would be on the order of nanometers, whereas errors in human height measurements are probably on the order of a centimeter or two) and partly on the quality of the tools being used. Physics and chemistry labs have highly accurate, expensive scales that can measure mass to the nearest gram, microgram, or nanogram, whereas the average scale in someone's bathroom is probably accurate within a half-pound or pound.

To wrap your head around random error, imagine that you are baking a cake that requires 6 tablespoons of butter. To get the 6 tablespoons of butter (three-quarters of a stick, if there are 4 sticks in a pound, as is usually true in the US), you could use the marks that appear on the waxed paper around the stick, assuming they are lined up correctly. Or you could perhaps follow my mother's method, which is to unwrap the stick, make a slight mark at what looks like one-half of the stick, and then get to three-quarters by eyeballing half of the one-half. Or you could use my method, which is to eyeball the three-quarter mark from the start and slice away. Any of these "measurement" methods will give you roughly 6 tablespoons of butter, which is certainly good enough for the purposes of baking a cake—but probably not exactly 3 ounces' worth, which is how much 6 tablespoons of butter weighs in the US.[i] The extent to which you're slightly over 3 ounces this time and perhaps slightly under 3 ounces next time is causing random error in your measurement of butter. If you always underestimated or always overestimated, then that would be a bias—however, your consistently under- or overestimated measurements would within themselves contain random error.

Inherent Variability

For any given variable that we might want to measure in epidemiology (e.g., height, GPA, heart rate, number of years working at a particular factory, serum triglyceride level, etc.), we expect there to be variability in the sample—that is, we do not expect everyone in the population to have exactly the same value. This is not random error. Random error (and bias) occurs when we try to *measure* these things. Indeed, epidemiology as a field relies on this inherent variability. If everyone were exactly the same, then we would not be able to identify which kinds of people were at higher risk for developing a particular disease.

In epidemiology, sometimes our measurements rely on a human other than the study participant measuring something on or about the participant. Examples would include measured height or weight, blood pressure, or serum cholesterol. For some of these (e.g., weight and serum cholesterol), the random error creeps into the data because of the instrument being used—here, a scale that has probably a half-pound fluctuation, or a laboratory assay with a margin of error of a few milligrams per deciliter. For other measurements (e.g., height and blood pressure), the measurer themselves is responsible for any random error, as in the butter example.

However, many of our measurements rely on participant self-reporting. There are whole textbooks and classes devoted to questionnaire design, and the science behind how to get the most accurate data from people via survey methods is quite good. The Pew Research Center offers a nice introductory tutorial on questionnaire design on its website.

Relevant to our discussion here, random error will appear in questionnaire data as well. For some variables, there will be less random error than others (e.g., self-reported race is probably quite accurate), but there will still be some—for example, people accidentally checking the wrong box. For other variables, there will be more random error (e.g., imprecise answers to questions such as, "In the last year, how many times per month did you eat rice?"). A good question to ask yourself when considering the amount of random error that might be in a variable derived from a questionnaire is, "*Can* people tell me this?" Most people could theoretically tell you how much sleep they got last night, but they would be hard-pressed to tell you how much sleep they got on the same night one year ago. Whether or not they *will* tell you is a different matter and touches on bias (see chapter 6). Regardless, random error in questionnaire data increases as the likelihood that people *could* tell you the answer decreases.

Quantifying Random Error

While we can—and should—work to minimize random error (using high-quality instruments, training staff on how to take measurements, designing good questionnaires, etc.), it can never be eliminated entirely. Luckily, we can use statistics to quantify the random errors present in a study. Indeed, this is what statistics is for. In this book, I will cover only a small slice of the vast field of statistics: interpretation of ***p*-values** and **confidence intervals (CI)**. Rather than focus on how to calculate them[1], I will instead focus on what they mean (and what they do not mean). Knowledge of p-values and CIs is sufficient to allow accurate interpretation of the results of epidemiologic studies for beginning epidemiology students.

p-values

When conducting scientific research of any kind, including epidemiology, one begins with a hypothesis, which is then tested as the study is conducted. For example, if we are studying average height of undergraduate students, our hypothesis (usually indicated by H_1) might be that male students are, on average, taller than female students. However, for statistical testing purposes, we must rephrase our hypothesis as a **null hypothesis**[2]. In this case, our null hypothesis (usually indicated by H_0) would be the following:

1. There isn't just one formula for calculating a *p*-value or a CI. Rather, the formulas change depending on which statistical test is being applied. Any introductory biostatistics text that discusses which statistical methods to use and when would also provide the corresponding information on *p*-value and CI calculation.
2. Don't spend too long trying to figure out why we need a null hypothesis; we just do. The rationale is buried in centuries of academic philosophy of science arguments.

H_0: There is no difference in mean height between male and female undergraduate students.

We would then undertake our study to test this hypothesis. We first determine the target population (undergraduate students) and draw a sample from this population. We then measure the heights and genders of everyone in the sample, and calculate mean height among men versus that among women. We would then conduct a statistical test to compare the mean heights in the 2 groups. Because we have a continuous variable (height) measured in 2 groups (men and women), we would use a **t-test**[3], and the *t*-statistic calculated via this test would have a corresponding *p*-value, which is what we really care about.

A *p*-value is the probability that if you repeated the study, you would find a result at least as extreme, assuming the null hypothesis is true.

Let's say that in our study we find that male students average 5 feet 10 inches, and among female students the mean height is 5 feet 6 inches (for a difference of 4 inches), and we calculate a *p*-value of 0.04. This means that if there really is no difference in average height between male students and female students (i.e., if the null hypothesis is true) and we repeat the study (all the way back to drawing a new sample from the population), there is a 4% chance that we will again find a difference in mean height of 4 inches or more.

There are several implications that stem from the above paragraph. First, in epidemiology we always calculate 2-tailed *p*-values. Here this simply means that the 4% chance of a ≥4 inch height difference says nothing about which group is taller—just that one group (either males or females) will be taller on average by at least 4 inches. Second, p-values are meaningless if you happen to be able to enroll the entire population in your study. As an example, say our research question pertains to students in Public Health 425 (H425, Foundations of Epidemiology) during the 2020 winter term at Oregon State University (OSU). Are men or women taller in this population? As the population is quite small and all members are easily identified, we can enroll everyone instead of having to rely on a sample. There will still be random error in the measurement of height, but we no longer use a *p*-value to quantify it. This is because if we were to repeat the study, we would find exactly the same thing, since we actually measured everyone in the population. P-values only apply if we are working with samples.

3. How to choose the correct test is beyond the scope of this book—see any book on introductory biostatistics

Finally, note that the *p*-value describes the probability of your data, assuming the null hypothesis is true—it does not describe the probability of the null hypothesis being true given your data. This is a common interpretation mistake made by both beginning and senior readers of epidemiologic studies. The *p*-value says *nothing* about how likely it is that the null hypothesis is true (and thus on the flip side, about the truth of your actual hypothesis). Rather, it quantifies the likelihood of getting the data that you got if the null hypothesis *did* happen to be true. This is a subtle distinction but a very important one.

Statistical Significance

What happens next? We have a *p*-value, which tells us the chance of getting our data given the null hypothesis. But what does that actually mean in terms of what to conclude about a study's results? In public health and clinical research, the standard practice is to use $p \le 0.05$ to indicate **statistical significance.** In other words, decades of researchers in this field have collectively decided that if the chance of committing a **type I error** (more on that below) is 5% or less, we will "reject the null hypothesis." Continuing height example from above, we would thus conclude that there is a difference in height between genders, at least among undergraduate students. For *p*-values above 0.05, we "fail to reject the null hypothesis," and instead conclude that our data provided no evidence that there was a difference in height between male and female undergraduate students.

Failing to Reject the Null vs. Accepting the Null

If $p > 0.05$, we fail to reject the null hypothesis. We do not ever accept the null hypothesis because it is very difficult to prove the absence of something. "Accepting" the null hypothesis implies that we have proven that there really is no difference in height between male and female students, which is not what happened. If $p > 0.05$, it merely means that we did not find evidence in opposition to the null hypothesis—*not* that said evidence doesn't exist. We might have gotten a weird sample, we might have had too small a sample, etc. There is a whole field of clinical research (comparative effectiveness research[vi]) dedicated to showing that one treatment is no better or worse than another; the field's methods are complex, and the sample sizes required are quite large. For most epidemiologic studies, we simply stick to failing to reject.

Is the $p \le 0.05$ cutoff arbitrary? Absolutely. This is worth keeping in mind, particularly for *p*-values very near this cutoff. Is 0.49 really that different from 0.51? Likely not, but they are on opposite sides of that arbitrary line. The size of a *p*-value depends on 3 things: the sample size, the effect size (it is easier to reject the null hypothesis if the true difference in height—were we to measure everyone in the population, rather than only our sample—is 6 inches rather than 2 inches), and the consistency of the data, most commonly measured by the standard deviations around the mean heights in the 2 groups. Thus a *p*-value of 0.51 could almost certainly be made smaller by simply enrolling more people in the study (this pertains to **power**, which is the inverse of **type II error**, discussed below). It is important to keep this fact in mind when you read studies.

Frequentist versus Bayesian Statistics

Statistical significance testing is part of a branch of statistics referred to as *frequentist statistics.*[ii] Though extremely common in epidemiology and related fields, this practice is not generally regarded as an ideal science, for a number of reasons. First and foremost, the 0.05 cutoff is entirely arbitrary,[iii] and strict significance testing would reject the null for p = 0.049 but fail to reject for p = 0.051, even though they are nearly identical. Second, there are many more nuances to interpretation of *p*-values and confidence intervals than those I have covered in this chapter.[iv] For instance, the *p*-value is really testing all analysis assumptions, not just the null hypothesis, and a large *p*-value often indicates merely that the data cannot discriminate among numerous competing hypotheses. However, since public health and clinical medicine both require yes-or-no decisions (Should we spend resources on that health education campaign? Should this patient get this medication?), there needs to be *some* system for deciding yay or nay, and statistical significance testing is currently it. There are other ways of quantifying random error, and indeed Bayesian statistics (which instead of a yes-or-no answer yields a probability of something happening)[ii] is becoming more and more popular. Nonetheless, as frequentist statistics and null hypothesis testing are still by far the most common methods used in epidemiologic literature, they are the focus of this chapter.

Type I and Type II errors

A type I error (usually symbolized by α, the Greek letter *alpha*, and closely related to *p*-values) is the probability that you incorrectly reject the null hypothesis – in other words, that you "find" something that's not really there. By choosing 0.05 as our statistical significance cut-off, we in the public health and clinical research fields have tacitly agreed that we are willing to accept that 5% of our findings will really be type I errors, or *false positives*.

A type II error (usually symbolized by β, the Greek letter *beta*) is the opposite: β is the probability that you incorrectly fail to reject the null hypothesis—in other words, you miss something that really is there.

Power = 1 – β and is interpreted as the likelihood that you'll find things if they are there.

Power in epidemiologic studies varies widely: ideally it should be at least 90% (meaning the type II error rate is 10%), but often it is much lower. Power is proportional to sample size but in an exponential manner—power goes up as sample size goes up, but to get from 90 to 95% power requires a much larger jump in sample size than to go from 40 to 45% power. If a study fails to reject the null hypothesis, but the data look like there might be a large difference between groups, often the issue is that the study was underpowered, and with a larger sample, the *p*-value would probably fall below the magic 0.05 cutoff. On the other hand, part of the issue with small samples is that you might just by chance have gotten a non-representative sample, and adding additional participants would *not* drive the results toward statistical significance. As an example, suppose we are again interested in gender-based height differences, but this time only among collegiate athletes. We begin with a very small study—just one men's team and one women's team. If we happen to choose, say, the men's basketball team and the women's gymnastics team, we are likely to find a whopping difference in mean heights—perhaps 18 inches or more. Adding other teams to our study would almost certainly result in a much narrower difference in mean heights, and the 18 inch difference "found" in our initial small study would not hold up over time.

Confidence Intervals

Because we have set the acceptable α level at 5%, in epidemiology and related fields, we most commonly use 95% confidence intervals (95% CI). One can use a 95% CI to do significance testing: if the 95% CI does not include the null value (0 for risk difference and 1.0 for odds ratios, risk ratios, and rate ratios), then $p < 0.05$, and the result is *statistically significant.*

Though 95% CI can be used for significance testing, they contain much more information than just whether the p-value is <0.05 or not. Most epidemiologic studies report 95% CI around any **point estimates** that are presented. The correct interpretation of a 95% CI is as follows:

> If you repeated the study 100 times (back to drawing your sample from the population), and the study is free of all bias, then 95 of those 100 times the CI that you calculate would include the "real" answer that you would get were you able to enroll everyone in the population.

We can also illustrate this visually:

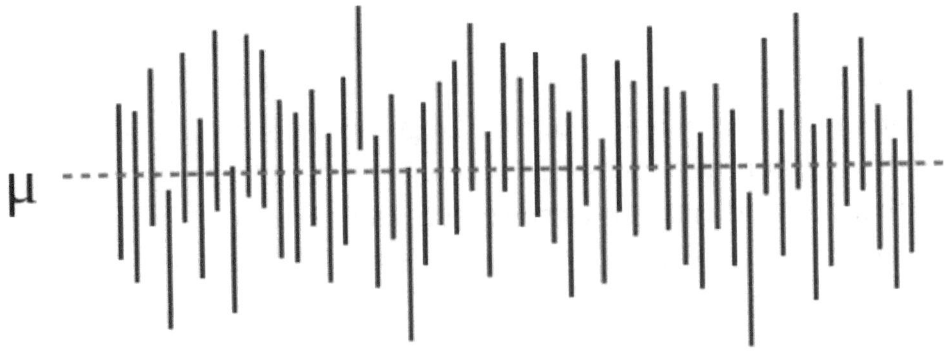

Figure 5-1
Source: *https://es.wikipedia.org/wiki/Intervalo_de_confianza*

In Figure 5-1, the population parameter μ represents the "real" answer that you would get if you could enroll absolutely everyone in the population in the study. We estimate μ with data from our sample. Continuing with our height example, this might be 5 inches: if we could magically measure the heights of every single undergraduate student in the US (or the world, depending on how you defined your target population), the mean difference between male and female students would be 5 inches. Importantly, this population parameter is almost always unobservable—it only becomes observable if you define your population narrowly enough that you can enroll everyone. Each blue vertical line represents the CI of an individual "study"—50 of them, in this case. The CIs vary because the sample is slightly different each time—however, most of the CIs (all but 3, in fact) do contain μ.

If we conduct our study and find a mean difference of 4 inches (95% CI, 1.5 – 7), the CI tells us 2 things. First, the *p*-value for our *t*-test would be <0.05, since the CI excludes 0 (the null value in this case, as we are calculating a difference measure). Second, the interpretation of the CI is: if we repeated our study (including drawing a new sample) 100 times, then 95 of those times our CI would include the real value (which we know here is 5 inches, but which in real life you would not know). Thus looking at the CI here of 1.5 – 7.0 inches gives an idea of what the real difference might be—it almost certainly lies somewhere within that range but could be as small as 1.5 inches or as large as 7 inches. Like *p*-values, CIs depend on sample size. A large sample will yield a comparatively narrower CI. Narrower CIs are considered to be better because they yield a more precise estimate of what the "true" answer might be.

Summary

Random error is present in all measurements, though some variables are more prone to it than others. P-values and CIs are used to quantify random error. A p-value of 0.05 or less is usually taken to be "statistically significant," and the corresponding CI would exclude the null value. CIs are useful for expressing the potential range of the "real" population-level value being estimated.

References

i. Butter in the US and the rest of the world. *Errens Kitchen.* March 2014. https://www.errenskitchen.com/cooking-conversions/butter-measurement-weight-conversions/. Accessed September 26, 2018. (↲ Return)

ii. Bayesian vs frequentist approach: same data, opposite results. *365 Data Sci.* August 2017. https://365datascience.com/bayesian-vs-frequentist-approach/. Accessed October 17, 2018. (↲ Return 1) (↲ Return 2)

iii. Smith RJ. The continuing misuse of null hypothesis significance testing in biological anthropology. *Am J Phys Anthropol.* 2018;166(1):236-245. doi:10.1002/ajpa.23399 (↲ Return)

iv. Farland LV, Correia KF, Wise LA, Williams PL, Ginsburg ES, Missmer SA. P-values and reproductive health: what can clinical researchers learn from the American Statistical Association? *Hum Reprod Oxf Engl.* 2016;31(11):2406-2410. doi:10.1093/humrep/dew192 (↲ Return)

v. Greenland S, Senn SJ, Rothman KJ, et al. Statistical tests, p values, confidence intervals, and power: a guide to misinterpretations. *Eur J Epidemiol.* 2016;31:337-350. doi:10.1007/s10654-016-0149-3

vi. Why is comparative effectiveness research important? Patient-Centered Outcomes Research Institute. https://www.pcori.org/files/why-comparative-effectiveness-research-important. Accessed October 17, 2018. (↲ Return)

6. Bias

Learning Objectives

After reading this chapter, you will be able to do the following:

1. Define *bias*, and differentiate it from random error
2. Differentiate between the different types of bias common to epidemiologic studies, and provide illustrative examples of each

As we learned in the previous chapter, **random error** exists in all studies, because it exists to some degree in all measurements. Standard statistical methods are used to quantify random error and the role it may or may not have played in the interpretation of a study's results. Random errors cannot be eliminated entirely, and by correctly interpreting p-values and confidence intervals (CIs), we can place our results in the appropriate context.

Bias, on the other hand, refers to systematic errors, meaning that they disproportionately affect the data in one direction only—so, for example, we would always underestimate or always overestimate when cutting the 6 tablespoons of butter for our cake (see chapter 5). There are many potential sources of bias in epidemiologic studies; here we will cover some of the most common. As with random error, all studies contain some degree of bias, and like with random error, we do our best to minimize it. The difference is that statistical methods cannot help us with bias.

Bias results in a calculated **measure of association** that is either above or below what it "should" be, because our data were skewed in one direction or the other. It is impossible to know the magnitude of the bias or even the direction. Did we over- or underestimate the risk ratio (RR)? By how much? We will never know the answers to these questions, but by thinking through likely directions of systematic errors (e.g., people will typically overestimate how much exercise they get), we can often make educated guesses about the direction of a bias and perhaps also its magnitude. But they are only guesses.

Bias can be minimized with correct study design and measurement techniques, but it can never be omitted entirely. All studies have bias because humans are involved, and humans are inherently biased.[i] Good scientists will ponder potential sources of bias during study planning phases, working to minimize them. They will also make an honest appraisal of residual bias at the end of a study and discuss this in the limitations section of a paper's discussion section (see appendix 1).

Internal versus External Validity

Bias can affect both the **internal validity** and the **external validity** of a study. The former is a much more serious issue. Internal validity refers to the inner workings of a study: Was the best design used? Were variables measured in a reasonable way? Did the authors conduct the correct set of analyses? Note that although we can't measure or quantify internal validity, an understanding of epidemiology and biostatistics allows for a qualitative appraisal. We can believe the results of a study that appears to be internally valid. A study that has major methodologic issues, however, lacks internal validity, and we probably should not accept the results.

If a study lacks internal validity, stop. There is rarely a need to assess it further. On the other hand, if a study does seem to have internal validity, we then assess external validity, or **generalizability**. External validity refers to how well the results of this particular study could be applied to the larger population. Recall from chapter 1 that the **target population** is the group about whom we wish to say something, using data collected from our sample. Occasionally, we find a study that is internally valid—meaning, it was conducted in an entirely correct way—but for some reason, the **sample** is not sufficiently representative of the target population. For example, during my dissertation work, I used cohort data to estimate the effects of maternal physical activity during pregnancy on various birth outcomes.[ii] The data came from a large pregnancy cohort and included data on hundreds of exposures and dozens of outcomes.[iii] The inclusion criteria were lenient—all women pregnant with a **singleton fetus** planning to deliver at a certain hospital were eligible. Much like we find in the general population, the pregnant people in this cohort were mostly sedentary.[ii,iv]

In some more recent work, I was looking specifically at physical activity during pregnancy as the only exposure; thus, my advertisement to recruit women into the study mentioned that I was studying exercise in pregnancy (rather than pregnancy in general).[1] In this more recent study, I had very few sedentary people—indeed, I have a few who reported running half marathons while pregnant! Since this is not normal, my study—though it does have reasonable internal validity—cannot be generalized to all pregnant women but only to the subpopulation of them who get a fair bit of physical activity. It lacks *external validity.* Because it has good internal validity, I can generalize the results to highly active pregnant women—just not to all pregnant women.

1. All research with human participants must be approved by an ethics board, usually called an *institutional review board* (IRB) in the US. The IRB must approve all study materials, including advertisements, and all such advertisements must clearly state the research question.

Representativeness of Samples

If you have a sample that is not representative of the underlying population, this affects external validity. The extent to which this is a concern, though, depends on the research question. Questions that apply mainly to biology (e.g., do statins lower serum cholesterol levels?) do not necessarily require representative samples, because physiology does not usually vary to any great extent between people with different demographic characteristics (differences by sex are the one exception): my body likely processes statin drugs in a nearly-identical manner to that of most other women's. However, when the research question involves behavior, then we must be very concerned about representativeness, because behavior varies greatly by demographics and social context. Thus, "Do statins lower serum cholesterol levels?" is a very different question than "If you prescribe statins for people with high cholesterol, will they live longer?" since the latter requires behavior on both the clinician's part (providing the prescription) and the patient's part (filling the prescription and then taking the medication as directed).

Figure 6-1 illustrates the difference between internal validity and external validity. This uses the cohort diagram, but the same principle applies to all study designs:

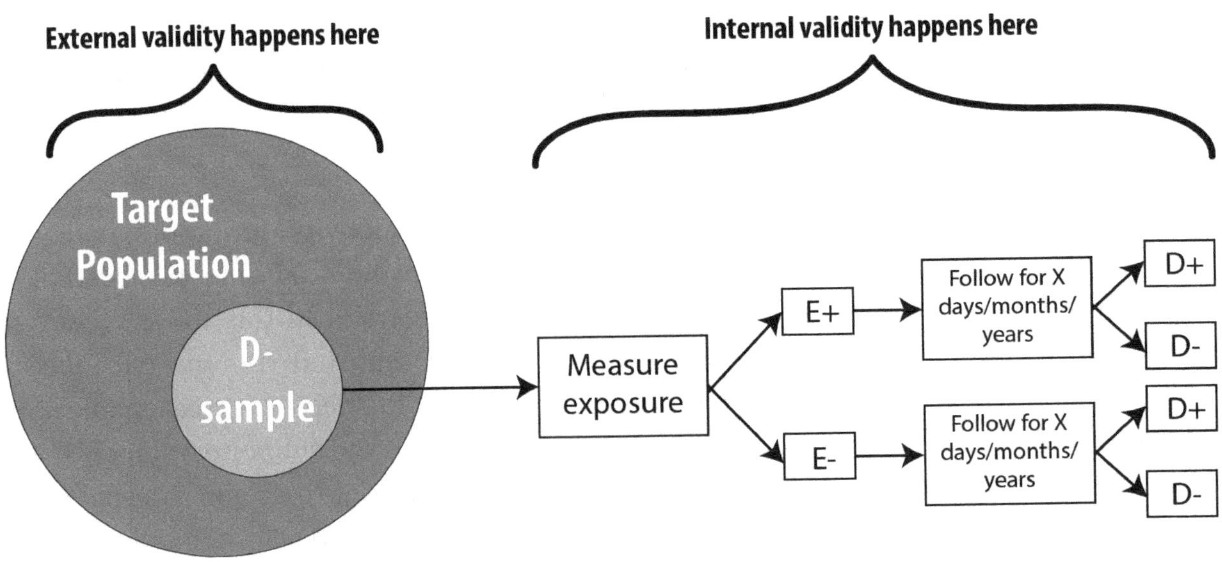

Figure 6-1

Selection Bias

Selection bias can affect either the internal or the external validity of a study. The above example about exercise in pregnancy (where I had a non-representative sample from the population) is the kind of selection bias affecting external validity: my results are generalizable only to the subset of the population from whom my sample actually was drawn rather than to the entire population. This sort of selection bias is not ideal, but one can easily recover by simply narrowing the target population to whom the results will apply. Asking "Who did the researchers get? Who did they miss?" will help in assessing the extent of overall selection bias.

In other cases, selection bias can affect internal validity. This is much worse, since the results of that study cannot be applied to anyone, because it has fundamental flaws. Selection bias adversely affecting internal validity occurs when the exposed and unexposed groups (for a cohort study) or the diseased and nondiseased groups (for a case-control study) are not drawn from the same population. For example, in a study of maternal physical activity and labor outcomes,[vi] the "active" group was recruited from a prenatal exercise class, but the "sedentary" group was recruited from prenatal care clinics. To the extent that people who voluntarily choose to pay for and attend an exercise class specifically for pregnant women are different than the overall group of women getting prenatal care, this study has a selection bias affecting internal validity, because the exposed and unexposed groups (samples) come from different populations. Again, asking, "Who did they get? Who did they miss?" and also "Was this different between the two groups?" will help here.

Selection bias affecting internal validity can creep up in other, less obvious ways, mostly relating to missing data. Was the participation rate different between the 2 groups? Was there more loss to follow-up in one group versus the other? Either of these could lead to groups that might not reflect the same underlying population, since the kinds of people who agree to participate in studies are different than those who don't, and the kinds of people who drop out are different than those who don't. For instance, in studies of older adults, the sickest patients tend to drop out because they become too sick to attend the study-related clinic visits. If this occurs more in one study group than the other, it leads to selection bias.

Healthy worker bias is a type of selection bias, and it refers to the fact that people who can work are generally healthier than the overall population because the overall population includes people who are too sick to work. Thus studies that recruit from a population of people who work may lack external generalizability—which is fine, as long as one is careful when applying the study's results. However, healthy worker bias can also affect internal validity if one group is recruited specifically from a population of workers and the other from the general population. For instance, if we suspect that Factory A has an environmental toxin (and our cohort study's exposed group consists of workers from Factory A), then our unexposed group needs to be workers from somewhere else—not, say, spouses or neighbors (who may or may not work) of the exposed participants.

Misclassification Bias

Misclassification refers simply to measuring things incorrectly, such that study participants get put into the wrong box in the 2 x 2 table: we call them "diseased" when really they're not (or vice versa); we call them "exposed" when really they're not (or vice versa).

Continuing with our **exercise** in pregnancy example, say we recruit 1,000 pregnant women and assess their levels of physical activity. We decide that anyone meeting the recommendation for physical activity during pregnancy (30 minutes of moderate activity, most days of the week[vii2]) will be classified as "exposed," and anyone reporting less activity will be "unexposed." In general, all people will over-report their levels of physical activity.[ix(p46)] Thus in our study of 1,000 women, we would expect some level of misclassification—if everyone slightly overreports their amount of physical activity, then those people who actually got just *under* the recommended amount will be incorrectly classified because their over-reporting will bump them up into the exposed (met the guidelines) group.

If this is what the data *should* look like (imagine any disease you like here):

	D+	D-
E+	200	100
E-	300	400

But instead, if we incorrectly classify some women as exposed because of overreporting, the table might look like this:

	D+	D-
E+	230	140
E-	270	360

This is called **nondifferential misclassification**, because it occurs at the same rate (here, 10% of

2. The astute among you will notice that these recommendations look remarkably like the recommendations[viii] for non-pregnant persons. Indeed, barring certain well-defined and relatively rare complications, pregnant women should be just as active as non-pregnant people.

unexposed were incorrectly classified as exposed) in both the diseased and nondiseased groups. Nondifferential misclassification is not quite the same as random error—in random error, we might have 10% misclassification, but it would go in both directions. Here we really only expect overreporting of physical activity, so it is a systematic error, or bias. Misclassification, like all other forms of bias, affects studies by giving us the wrong estimate of association.

Misclassification example

Using the first 2 x 2 table above (ie, the "correct" data—note that this is almost never observable), the odds ratio (OR) is:

$$OR = \frac{200 \times 400}{300 \times 100} = 2.67$$

Whereas the odds ratio for the biased data (the ones we actually collected in our study) is:

$$OR = \frac{230 \times 360}{140 \times 270} = 2.19$$

The result of the calculation with data including nondifferential misclassification is closer to the null than the correct one would be. In real life, we cannot ever observe the "correct" table, and thus we cannot know by how much or in which direction our estimate is biased—just that it is.

Nondifferential Misclassification: Bias towards the Null?

In the exercise in pregnancy example, the OR estimate based on the biased data was biased toward the null (i.e., it's closer to 1.0, the null value for odds ratios), but it could just as easily have been biased away from the null. Some older epidemiology textbooks will say that nondifferential misclassification always biases toward the null, but it turns out that this is not true.[9p143] It's best to assume that you don't know which way the bias is going.

On the plus side, even if the data are misclassified, as long as it's nondifferential misclassification, we have probably still ranked people correctly. If *everyone* overestimates their physical activity, we can still tell the couch potatoes from the marathon runners. Thus, although the estimate of association we calculated with our misclassified data (in which everyone added, say, 30–60 minutes to their weekly exercise totals) is almost certainly not "correct," with nondifferential misclassification, we can often still say something about the results (perhaps that the more exercise one gets, the lower one's risk of heart disease?). This statement will almost certainly remain true, even if we were able to correct for the overestimate of everyone's physical activity and generate an unbiased estimate of association.

The flip side is **differential misclassification.** With differential misclassification, we again find that some people are put into the wrong boxes in the 2 × 2 table, but this time it is not equally distributed across all study groups. Perhaps diseased people misreport more than nondiseased people. Or perhaps investigators are subconsciously more likely to classify someone as "diseased" if they are known to be exposed. Differential misclassification is considered a fatal threat to a study's internal validity. Study authors, knowing this, will often acknowledge measurement errors in their study but claim that they are nondifferential and therefore essentially irrelevant. When you encounter such claims, think it through for yourself, and be sure you agree with the authors before citing their work. Differential misclassification is more common than many of us would like to admit.

Misclassification goes by numerous other names, including social desirability bias, interviewer bias, clinician bias, recall bias, and so on. Regardless of name, however, misclassification boils down to people being called exposed when they're not, not exposed when they are, not diseased when really they are, or diseased when really they're not. When considering self-reported data, as discussed in the previous chapter, you must first ask yourself, "*Can* people tell me this?" If not, stop. But if yes, then you must consider, "Will people tell me this?" If not, then the data may have bias from misclassification.

Sensitivity Analyses

Sometimes called *bias analysis*, a *sensitivity analysis* is a set of extra analyses conducted after the main results of a study are known, with the goal of quantifying how much bias there might have been and in which direction it shifted the results. Not all research questions and datasets are amenable to sensitivity analysis, but for those that are, it's a great way for authors to increase the perceived validity of their results. There is no set way of conducting a sensitivity analysis; rather, one examines all assumptions made as part of an analysis and tests the extent to which those assumptions, rather than an underlying true association, drove the results.

For example, if we were studying physical activity, and in our main analysis decided that anyone meeting the guidelines was "active" and all other people in the study were "sedentary," then one sensitivity analysis might change this cutoff point (perhaps now we declare that anyone accumulating 2 or more hours per week of exercise is "active," even though this is less than the guidelines suggest) and see what the new estimate of association is. If the new estimate is close to the original one, then we can conclude that our choice of cutoff point (an assumption we made during analysis) did not affect the results extensively. This alone does not preclude the possibility that the original results are incorrect, but it does lessen the possibility that we would find a vastly different answer if we repeated the study using slightly different methods.

Missing data on individual variables also leads to misclassification—for instance, in the US, people do not like to talk about money, so often questions on income go unanswered. If the kind of person who leaves the income question blank is different than the kind of person who answers it, then the data are not **missing at random**. If data truly *are* missing at random (which might happen if, for instance, some people genuinely don't see the question because of a quirk in the page layout), then the result is a slightly smaller sample size (and correspondingly less **power**), but otherwise this has no adverse effects. However, in real life, data are almost never missing at random, which means they are missing according to some pattern—and thus are creating a bias. If study authors claim that they have data missing at random, think carefully about the scenario and make sure you agree. More often, study authors simply don't mention missing data at all.[3] If an important variable in the analysis is missing for more than 5% of the sample, yet this is not discussed by the authors, then be wary of the results. They are probably biased.

Publication Bias

Publication bias arises because papers with more exciting results are more likely to get published. A paper whose main finding is "there is no association between x and y" is difficult to get published—so much so that there is an entire, legitimate, peer-reviewed journal dedicated solely to publishing these so-called negative results.

This type of bias does not apply to individual studies, but rather to areas of the literature as a whole. If papers with larger estimates of association and/or smaller *p*-values are more likely to get published, then when you attempt to look at the entire body of literature on a given topic (e.g., should elderly people take prophylactic aspirin to prevent heart attacks?), the picture you get is biased, because only the exciting papers were published. All the papers that showed no effect of aspirin on heart attack were not published. This is worth keeping in mind whenever you are doing literature searches and is discussed further in chapter 9.

Conclusion

All epidemiologic studies include bias. Investigators can minimize the biases that are present through good design and measurement methods, but some will always remain. Those biases affecting a study's internal validity (selection bias that pertains more to one group than another, or differential misclassification) render that study either entirely useless or useful only with extreme caution. Selection bias affecting external validity only—the presence of nondifferential

3. Strange but true!

misclassification or selection bias operating on the entire sample,—is manageable as long as one understands the associated limitations. Missing data, and the extent to which non-participation or non-compliance might have affected the results, should always be considered carefully.

References

i. Johnson CY. Everyone is biased: Harvard professor's work reveals we barely know our own minds. *Boston Globe*. 2013. https://www.boston.com/news/science/2013/02/05/everyone-is-biased-harvard-professors-work-reveals-we-barely-know-our-own-minds. Accessed November 27, 2018. (↲ Return)

ii. Bovbjerg M, Siega-Riz A, Evenson K, Goodnight W. Exposure assessment methods affect associations between maternal physical activity and cesarean delivery. J *Phys Act Health*. 2015;12(1):37-47. (↲ Return 1) (↲ Return 2)

iii. Pregnancy, Infection, and Nutrition (PIN). UNC Gillings School of Global Public Health. https://sph.unc.edu/epid/pregnancy-infection-and-nutrition-pin/. Accessed October 18, 2018.

iv. Evenson KR, Savitz DA, Huston SL. Leisure-time physical activity among pregnant women in the US. *Paediatr Perinat Epidemiol*. 2004;18(6):400-407. doi:10.1111/j.1365-3016.2004.00595.x (↲ Return)

v. Rothman KJ, Gallacher JEJ, Hatch EE. Why representativeness should be avoided. *Int J Epidemiol*. 2013;42(4):1012-1014. doi:10.1093/ije/dys223

vi. Beckmann CR, Beckmann CA. Effect of a structured antepartum exercise program on pregnancy and labor outcome in primiparas. J *Reprod Med*. 1990;35(7):704-709. (↲ Return)

vii. ACOG committee opinion. Exercise during pregnancy and the postpartum period. 2002. American College of Obstetricians and Gynecologists. *Int J Gynaecol Obstet Off Organ Int Fed Gynaecol Obstet*. 2002;77(1):79-81. (↲ Return)

viii. Physical activity for everyone: Guidelines. Centers for Disease Control and Prevention (CDC). http://www.cdc.gov/physicalactivity/everyone/guidelines/adults.html. Accessed February 10, 2014. (↲ Return)

ix. Dishman RK, Heath GW, Lee I-M. *Physical Activity Epidemiology*. 2nd ed. Champaign, IL: Human Kinetics; 2013. (↲ Return)

7. Confounding

Learning Objectives

After reading this chapter, you will be able to do the following:

1. Explain the concept of confounding, and how it affects the results of epidemiologic studies
2. Reiterate the criteria that a variable must meet to be a possible confounder
3. Conduct a stratified analysis to determine whether a variable is a confounder
4. Provide examples of exposure/outcome/confounder relationships, in terms of confounder criteria and analysis requirements

Like random error and bias, **confounding** is another threat to study validity. Indeed, there are some texts,[i(p.37)] as well as papers,[ii] that refer to confounding as "confounding bias." I prefer the term *confounding*, without the word bias, because while it also leads to a systematic error in the data, confounding is a special case.

Imagine that you are doing a cross-sectional study in elementary school-aged kids of foot size and reading ability:

Figure 7-1

In words, does foot size affect reading ability?

You go to an elementary school and measure both foot size (measured as length in inches) and reading ability (measured in terms of words read per minute, averaged over a 5-minute testing period), and you collect the following data:

Table 7-1

Participant #	Foot Size (inches)	Reading Speed (wpm)
1	7.2	40
2	7.7	85
3	7.2	63
4	7.6	52
5	7.4	51
6	7.1	41
7	7.0	82
8	7.2	60
9	7.6	53
10	7.5	55
11	8.3	123
12	8.2	97
13	8.5	108
14	8.1	111
15	8.2	109
16	8.2	99
17	8.7	95
18	8.0	110
19	8.5	121
20	8.2	108
21	9.4	128
22	8.1	117
23	9.8	115
24	8.8	109
25	9.1	112
26	9.3	112
27	9.8	106
28	9.2	125
29	9.6	163
30	9.0	137

As discussed in chapter 4, in this book we will always dichotomize (i.e., split in two) continuous variables to make the math simpler. If we dichotomize both foot size and reading speed—at 8.25" and 100 wpm, respectively[1] —we can draw the following 2 x 2 table:

Table 7-2

		Reading Speed	
		<100	100+
Foot Size	<8.25"	12	5
	8.25"+	1	12

Because this is a cross-sectional study, we would calculate the odds ratio:

$$OR = \frac{AD}{BC} = \frac{(12)(12)}{(1)(5)} = 28.8$$

In words,

> Children with feet that are at least 8.25" long are 28.8 times as likely to be able to read at least 100 words per minute, compared to children with shorter feet.

Wow! This is a huge finding! Should we give all grade-school kids growth hormones so that they get bigger feet and increase their reading speeds?

Not so fast.

Given that the target population for this hypothetical study is grade-school children, it seems likely that there is a confounder at work—namely, grade level. Kids in higher grades will have bigger feet because they are older, and they will also by and large be faster readers:

1. The 8.25" and 100 wpm cutoffs would be a great thing to vary in a sensitivity analysis! See chapter 6.

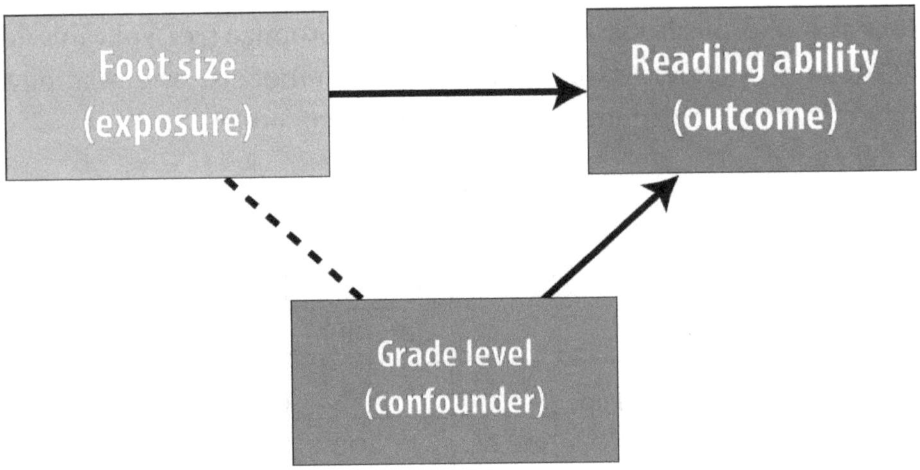

***Figure* 7-2**

In this scenario, we need to control for the confounder (grade level): we need to remove its influence to get an accurate estimate of the association between the exposure (foot size) and the outcome (reading ability).

Before we delve into how to control for confounders, let's discuss what confounders are from a theoretical perspective.

Criteria for Confounders

There are 3 criteria that a variable must meet in order for it to be a potential confounder (I say "potential" because not all variables that meet these criteria will actually turn out to confound the data—you figure this out during the analysis):

1. The variable must be statistically *associated with* the exposure.
2. The variable must *cause* the outcome.
3. The variable must *not* be on a causal pathway.

Let's discuss each of these in more detail.

Criterion #1: Associated with Exposure

Association is a statistical term that does not necessarily imply a causal relationship (this is discussed in more detail later, see chapter 10). Basically, association means that the confounding variable is more common in the exposed group than the unexposed group (or vice versa), thus producing a statistical association. The confounder does not need to cause or prevent the exposure, it just needs to be disproportionately distributed between the exposed and unexposed groups. In our previous example, grade level is disproportionately distributed among various foot sizes—kids in higher grades are more likely to have bigger feet compared to kids in lower grades. Note that there *can* be a causal relationship, with the confounder causing the exposure (but not the other way around—see criterion 3), but this is not necessary. In our example, grade level is not causing foot size (age is causing foot size)—but they are associated.

Criterion #2: Causes the Outcome

In this case, there must be a causal link between the confounder and the outcome. It does not have to be a proven causal link, just an "it is reasonably possible that this exposure causes (or prevents) that outcome" link. In our foot size/reading ability example, grade level (the confounder) certainly causes faster reading speed (the outcome).

Importantly, the confounder must cause the outcome—not the other way around. If the outcome is causing the confounder, then it's not a confounder. There are many times in epidemiology when we aren't sure which way a causal arrow would go—does the disease cause the confounder, or does the confounder cause the disease? An example might be excessive weight loss and illness. Losing a large amount of weight quickly can make one ill—but being ill can also cause a large amount of weight loss. In scenarios like this, where we aren't sure which way the arrow points, what epidemiologists do in practice is first assume the arrow goes one way and do the analysis accordingly (here, that would mean either including or not the potential confounder). They then assume the arrow goes the other way and do the analysis again. If the results of both analyses are similar, then the arrow direction isn't important. But if the 2 analyses produce very different results, then we would report both and let the reader decide which is more applicable for them.

Criterion #3: Not on the Causal Pathway

The final criterion for a variable to be a potential confounder is that it is not be on the causal pathway from exposure to outcome. In other words, we do not want this scenario:

Figure 7-3

An example of a variable on a causal pathway might be as follows:

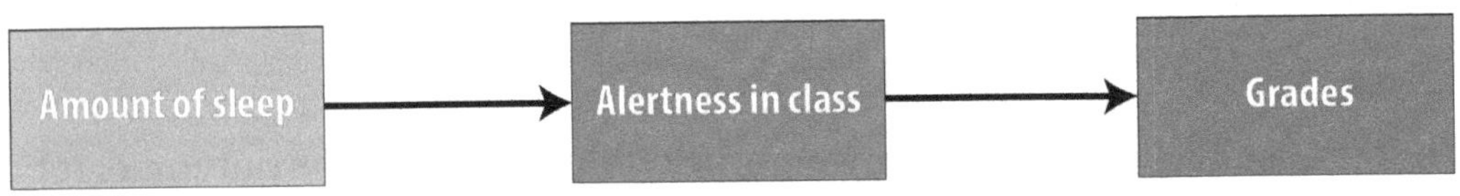

Figure 7-4

In this case, "alertness in class" is not a confounder, because it's *caused by* the amount of sleep and is thus on the causal pathway. Variables on the causal pathway are *mediators*, not potential confounders.

Confounding: Definition

A confounder is thus a third variable—not the exposure, and not the outcome[2]—that biases the measure of association we calculate for the particular exposure/outcome pair.

Importantly, from a research perspective, we never want to report a measure of association that is confounded. Imagine if we do our cross-sectional study on foot size and reading ability, without

2. Just a reminder! When epidemiologists say *outcome* we mean "health outcome or disease under study"—we do *not* mean the results of a study. Those are *results*. See Appendix 1.

accounting for grade level. We would report the odds ratio of 28.8 as calculated above...Oops! We've reported an association that's not really true—it's just confounded by grade level.

Methods of Confounder Control

One can control for confounding through either study design or analytic techniques. In terms of study design, you can

1. Restrict the sample
2. Match on the confounder
3. Randomize (as in, choose a randomized controlled trial as the study design)

Restricting the sample means that you limit your study only to one level of the confounder (e.g., third graders only). Therefore the potentially confounding variable no longer meets the first criterion for confounders—it cannot be disproportionately distributed between exposed and unexposed because there is only one level of the confounder available. Thus all exposed participants are in third grade, as are all unexposed. Our causal diagram now looks like this:

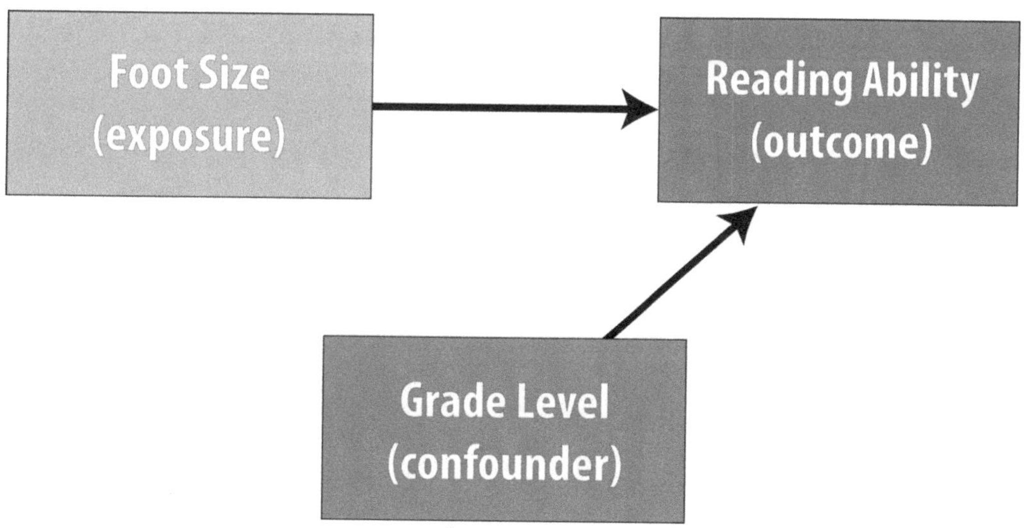

***Figure* 7-5**

By restricting to just one grade level, we remove the confounding by grade level: kids in both higher and lower grades are no longer relevant because if we have only third graders, then there *aren't* any kids in higher or lower grades. Among third-graders only, we would expect that foot size and reading ability are uncorrelated.

Inherent variability

Obviously not all third graders will have the same size feet, nor will all third graders uniformly have the same reading ability. However, on a group level, third graders *in general* have bigger feet and are better readers than first graders, and likewise have smaller feet and are poorer readers than fifth graders. Epidemiology as a science works because of both this individual variation and the fact that groups of people (selected on some characteristic, like grade level) are more similar to each other than they are to people in other groups.

Here are the same data with a column added for grade level:

Table 7-3

Participant #	Foot Size (inches)	Reading Speed (wpm)	Grade
1	7.2	40	1
2	7.7	85	1
3	7.2	63	1
4	7.6	52	1
5	7.4	51	1
6	7.1	41	1
7	7.0	82	1
8	7.2	60	1
9	7.6	53	1
10	7.5	55	1
11	8.3	123	3
12	8.2	97	3
13	8.5	108	3
14	8.1	111	3
15	8.2	109	3
16	8.2	99	3
17	8.7	95	3
18	8.0	110	3
19	8.5	121	3
20	8.2	108	3
21	9.4	128	5
22	8.1	117	5
23	9.8	115	5
24	8.8	109	5
25	9.1	112	5
26	9.3	112	5
27	9.8	106	5
28	9.2	125	5
29	9.6	163	5
30	9.0	137	5

Limiting ourselves to just third grade, then, the 2 x 2 table looks like this:

Table 7-4

		Reading Speed	
		<100	100+
Foot Size	<8.25″	2	2
	8.25″+	3	3

The odds ratio (OR) is 1.0. This is the correct measure of association to report. In reality, foot size has nothing to do with reading speed (OR 1.0). The 28.8 that we calculated earlier was wrong. It was confounded by grade level—there is no association once we control for this confounding by restricting to one grade level.

Though restriction works beautifully in terms of controlling confounding, often it is not a realistic approach because it limits our study too much. For instance, a reasonable study question might be, "What are predictors of breast cancer death among postmenopausal women?" Restricting by age (e.g., "What are predictors of breast cancer death among 62-year-old women?") would make the study much less useful because we wouldn't necessarily be able to generalize the results to women of other ages. Epidemiologists thus usually use other approaches for confounder control.

Matching is often used in case-control studies, and it has much the same effect as restriction in controlling confounding. For example, say we are looking at a particular birth defect (outcome) and maternal smoking (exposure), and we suspect that maternal age is a possible confounder. We would want to recruit a control with the same maternal age for each case: if the study were to enroll a 30-year-old case, we would want to match her with a 30-year-old control. The confounder (age) still causes the outcome (birth defects), but by forcing the confounder distribution to be the same between cases and controls, we have negated criterion #2, and thus negated the possible effect of the confounder on the exposure/outcome measure of association.

Randomizing works by forcing the confounder(s) to fail criterion # 1—in this case, by randomly assigning participants to the exposure, we have ensured an equal distribution of the confounder between the exposed and unexposed groups. The link between the confounder and the exposure is now missing, as it is with restriction. See chapter 9 for more on this.

In terms of controlling for confounding in the analysis phase, there are 2 main options:

1. Stratifying
2. Regression (which is really just a special case of stratifying)

To *stratify*, you take the data and make a different 2 x 2 table for each level of the potential confounder. Let's now assume that we are concerned with data from a case-control study on **oral contraceptive pill (OCP)** use (ever used vs. never used) and ovarian cancer:

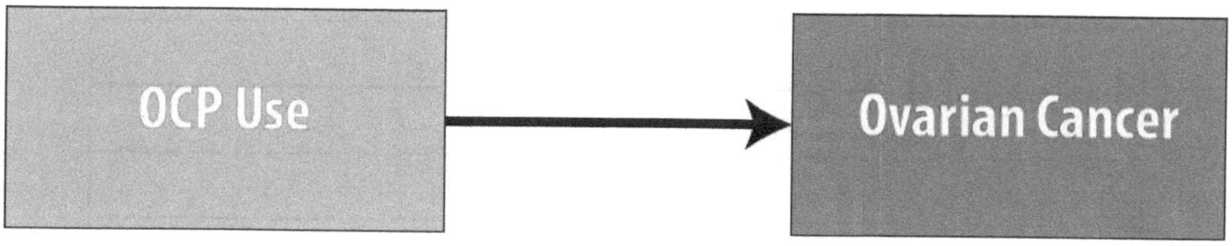

***Figure* 7-6**

We conduct this study and obtain the following data:

Table 7-5

Participant #	Ever OCP? 0 = no, 1 = yes	Ovarian Cancer? 0 = no (control), 1 = yes (case)
1	1	1
2	1	1
3	1	1
4	1	1
5	0	1
6	0	1
7	0	1
8	0	1
9	0	1
10	0	1
11	1	0
12	1	0
13	1	0
14	1	0
15	0	0
16	0	0
17	0	0
18	0	0
19	0	0
20	0	0

The 2 x 2 table would be as follows:

Table 7-6

		Ovarian Cancer	
		+	–
OCP	Ever	4	4
	Never	6	6

The OR is 1.0—use of oral contraceptives is not associated with ovarian cancer. During confounding analyses, this value is referred to as the crude or unadjusted measure of association—meaning that we have not yet accounted, adjusted, or controlled for any confounders. Unadjusted measures only take into account the exposure and the outcome.

However, what about smoking as a confounder? Let's check the confounder criteria:

1. The variable must be associated with the exposure.
    1. Yes! Both oral contraceptives and smoking increase one's risk of deep venous thrombosis, a potentially life-threatening condition. Smoking is thus considered a **contraindication** to oral contraceptive use,[iii] which leads clinicians to prescribe other forms of birth control instead for women who smoke. This leads to a disproportionate distribution of smokers (the confounder) between women who do and do not use oral contraceptives (the exposure).
2. The variable must cause the outcome.
    1. Possibly. While we often think of smoking as causing lung cancer (which it certainly does), smoking has also been associated with other cancers often enough that it is reasonable to suspect that it might cause ovarian cancer too.[iv]
3. The variable must not be on a causal pathway.
    1. Yes! It seems highly unlikely that taking birth control pills would in turn cause a woman to take up smoking.

Smoking thus meets our criteria and is a *potential* confounder in this scenario.[3] Here are the data with smoking status added:

3. Remember, variables that meet the confounder criteria are *potential* confounders. They may or may not actually produce a biased estimate of association; we figure this out during the analysis.

Table 7-7

Participant #	Ever OCP? 0 = no, 1 = yes	Ovarian Cancer? 0 = no (control), 1 = yes (case)	Smoker? 0 = no, 1 = yes
1	1	1	1
2	1	1	1
3	1	1	0
4	1	1	0
5	0	1	1
6	0	1	1
7	0	1	1
8	0	1	0
9	0	1	0
10	0	1	0
11	1	1	1
12	1	0	1
13	1	0	0
14	1	0	0
15	0	0	1
16	0	0	1
17	0	0	1
18	0	0	0
19	0	0	0
20	0	0	0

We now stratify by smoking status. In other words, we make 2 different 2 × 2 tables: one for smokers, and the other for nonsmokers. Keep in mind that all the women who appeared in the above 2 × 2 table for ovarian cancer and OCP use are still present—they're just in one of the two tables below, depending on whether they smoke or not.

Table 7-8

Smokers			
		Ovarian Cancer	
		+	–
OCP	Ever	2	3
	Never	3	2

Table 7-9

Nonsmokers			
		Ovarian Cancer	
		+	–
OCP	Ever	2	3
	Never	3	2

Note that the 2 x 2 tables are still for OCP (exposure) and ovarian cancer (outcome)—we have just made one such table for smokers and another for nonsmokers.

The next step in a stratified analysis is to calculate the ORs from these 2 x 2 tables, so we have an OR for smokers, and an OR for nonsmokers.

Confounding Example 1: OCP/Ovarian Cancer by Smoking Status

The odds ratio for smokers is:

$$OR_{smokers} = \frac{AD}{BC} = \frac{(2)(2)}{(3)(3)} = 0.44$$

Interpretation:

Women who have ovarian cancer are 0.44 times as likely to report a history of OCP use, compared to women without ovarian cancer—**among smokers only**.

And the odds ratio for non-smokers is:

$$OR_{non\text{-}smokers} = \frac{AD}{BC} = \frac{(2)(2)}{(3)(3)} = 0.44$$

Interpretation:

Among nonsmokers, women who have ovarian cancer are 0.44 times as likely to report a history of oral contraceptive (OCP) use, compared to women without ovarian cancer.

Note the additions (in red) to those interpretations! When conducting stratified analysis, it is important to say which group your measure of association applies to. This can come either at the beginning (as it does above for nonsmokers) or at the end (as it does above for smokers).

Since our stratum-specific odds ratios (0.44 for smokers and 0.44 for nonsmokers) are similar to each other but different from the crude OR (which was 1.0), we say that smoking is indeed acting as a confounder in these data. The crude OR was wrong; it was confounded by smoking.

"Similar" and "Different"—by How Much?

When the stratum-specific measures of association are similar to each other but different than the crude OR, we have confounding. But by how much? There is a standard criterion for "different"—if the crude and adjusted ORs are more than 10% different, most epidemiologists would consider that to be evidence of confounding. For "similar," though, there isn't really a consensus. Perhaps within 2–3% of each other? Importantly, the crude value does not fall between them.

The "real" OR is 0.44: oral contraceptive use is rather strongly associated with less ovarian cancer. But without accounting for smoking, it looks like this is not true (the crude OR was 1.0, and did not control for smoking). Because there is confounding, we thus would want to report an OR that controls for smoking (the confounder). The most common way to do this is to calculate an *"adjusted"* measure. There are many ways to calculate an adjusted measure of association[4]; one is to calculate the Mantel-Haenzel odds ratio:

$$\widehat{OR}_{MH} = \frac{\sum_{i=1}^{k}\left(\frac{a_i d_i}{n_i}\right)}{\sum_{i=1}^{k}\left(\frac{b_i c_i}{n_i}\right)}$$

$$\text{where } n_i = a_i + b_i + c_i + d_i$$

Figure 7-7
Source: *https://www.statsdirect.com/help/meta_analysis/mh.htm*

You can see from the formula that the Mantel-Haenzel OR is just a weighted average of the

4. Any biostatistics text would discuss several such methods.

stratum-specific odds ratios, with each stratum being an i. We call this the *adjusted* OR, and it has controlled for confounding by the variable on which we stratified.

Categorizing Continuous Variables

I mentioned in chapter 4 that if one has a continuous variable (e.g., age or height), most analyses are best served by keeping that variable continuous but that for the purposes of this book, we would dichotomize all variables to make the math easier. If a potential confounder is a continuous variable, we must categorize it (into 2 or 3 categories, usually) in order to conduct a stratified analysis by hand. Thus if height were our potential confounder, we might create 3 categories: less than 5′2″, 5′2″–6′0″, and taller than 6′0″. You can see, however, that within these categories, there is still considerable variability—5′2″ is a full 10″ shorter than 6′. Thus creating categories (i.e., strata for the stratified analysis) out of a continuous variable in order to control confounding might not work perfectly if the strata remain too heterogeneous. This produces *residual* confounding—we have removed some of the confounding by height but there is still some confounding left. This is one reason epidemiologists mostly jump straight to regression, in which it is easier to keep continuous variables continuous. However, the goal of this book is to create knowledgeable *readers* of epidemiologic studies, not knowledgeable *doers*, which would substantially more training. I thus rely on categorizing continuous variables so that the math is easy to follow and statistical software is unnecessary. If you follow the math as presented here, it is easy enough to make the cognitive leap to reading papers that use regression (regression is just stratified analysis with many more categories).

Most studies reported in the literature use the other method for controlling for confounding in analyses: *regression*, which is just a special case of a stratified analysis—specifically, it accounts for all possible strata. For instance, if we had continuous data on total months of smoking over the course of a lifetime, a regression model would "make" a 2 × 2 table for nonsmokers, then one for people had smoked for 1 month, then a 2 × 2 table for those who had smoked for 2 months, and so on, until each possible stratum had its own 2 × 2 table. The model then calculates a weighted average of the total of these (much like a mega-Mantel Haenzel), and the result is also known as the adjusted odds ratio. For cohort studies or randomized controlled trials we of course instead calculate the *adjusted risk ratio* or *adjusted rate ratio* (RR).

Interpretation

To interpret our OCP/ovarian cancer findings in words (the adjusted odds ratio, whether calculated via Mantel-Haenzel or by regression, is 0.44), we would say:

> Women who have ovarian cancer are 0.44 times as likely to report a history of OCP use compared to women without ovarian cancer, **controlling for smoking.**

Or we could say:

> Women who have ovarian cancer are 0.44 times as likely to report a history of OCP use compared to women without ovarian cancer, **adjusting for smoking.**

Or we could say:

> Women who have ovarian cancer are 0.44 times as likely to report a history of oral contraceptive use compared to women without ovarian cancer, **holding smoking constant.**

Notice how there are multiple ways of letting the reader know that smoking was treated as a confounder (**phrases in red**). It doesn't matter which you choose—the important thing is that you make it clear that we are presenting the measure of association having already dealt with the confounding.

Choosing Confounders

When conducting an analysis in real life, there are often multiple potential confounders. The first step in any analysis is to make a list of all such potential confounders. The easiest way to do this is first to make a list of all variables that might cause your outcome. Then take that list and make sure the variables are associated with the exposure. Finally, for any confounders that meet our first 2 criteria, make sure they are not on the causal pathway (e.g., that the exposure is not causing the confounder). As mentioned above, there are many instances where it is difficult to know which is causing which; in such cases, we do the analysis both ways.

The next step would be to determine which of the potential confounders meeting the 3 criteria to control for in an analysis (regression allows you to control for many confounders at once). One way would be to drop all confounders that do not meet the "10% change" criterion mentioned above. There are additional nuances, however, that are beyond the scope of this book, and prominent epidemiologists differ in their opinions on how to choose a list of confounders to control for.[v,vi,vii] Luckily, beginning epidemiology students will not need to conduct their own complex analyses; however, being able to think through a particular exposure/disease relationship and make a list of all potential confounders is a useful skill when reading the literature. Did the authors consider all the variables you thought of that meet the confounder criteria? If not, did they explicitly specify why not? If an obvious potential confounder is missing from the analysis in a particular article, then maybe that is not the most valid article.

Summary

Confounders are variables—not the exposure and not the outcome—that affect the data in undesirable and unpredictable ways. Specifically, in data that are confounded, one will calculate the wrong measure of association (and it is impossible to know in which direction one is wrong). This leads to inaccurate conclusions unless one controls for that confounder. To be a potential confounder, the variable must be statistically associated with the exposure, must cause the outcome, and must not be on the causal pathway. Potential confounders can be controlled for via study design (restriction, matching, or randomization) or during data analysis (stratification or regression, leading to an adjusted measure of association). In the latter case, if the crude and adjusted estimates of association are more than 10% different, the variable should be considered a confounder, and one would report the adjusted estimate because it controls for the confounder.

References

i. Last JM. A *Dictionary of Epidemiology*. 4th ed. 2001. New York: Oxford University Press. (↲ Return)

ii. Goldstein BA, Bhavsar NA, Phelan M, Pencina MJ. Controlling for informed presence bias due to the number of health encounters in an electronic health record. *Am J Epidemiol.* 2016;184(11):847-855. doi:10.1093/aje/kww112 (↲ Return)

iii. Bonnema RA, McNamara MC, Spencer AL. Contraception choices in women with underlying medical conditions. *Am Fam Physician*. 2010;82(6):621-628. (↲ Return)

iv. Study: Smoking causes almost half of deaths from 12 cancer types. American Cancer Society. https://www.cancer.org/latest-news/study-smoking-causes-almost-half-of-deaths-from-12-cancer-types.html. Accessed October 21, 2018. (↲ Return)

v. Greenland S, Pearl J, Robins JM. Causal diagrams for epidemiologic research. *Epidemiol Camb Mass*. 1999;10(1):37-48. (↲ Return)

vi. Harrell FEJ. *Regression Modeling Strategies, with Applications to Linear Models, Logistic Regression, and Survival Analysis*. New York: Springer; 2001. (↲ Return)

vii. Selvin S. *Statistical Analysis of Epidemiologic Data*. 3rd ed. Oxford: Oxford University Press; 2004. (↲ Return)

8. Effect Modification

Learning Objectives

After reading this chapter, you will be able to do the following:

1. Explain what effect modification is
2. Differentiate between confounders and effect modifiers
3. Conduct a stratified analysis to determine whether effect modification is present in the data

In the prior chapter, we discussed **confounding**. A confounder, you will recall, is a third variable that if not controlled appropriately, leads to a biased estimate of association. **Effect modification** also involves a third variable (not the exposure and not the outcome)—but in this case, we absolutely do not want to control for it. Rather, presence of effect modification is itself an interesting finding, and we highlight it.

When effect modification (also called *interaction*) is present, there will be different results for different levels of the third variable (also called a *covariable*). For example, if we do a cohort study on amount of sleep and GPA among Oregon State University (OSU) students over the course of one term, we might collect these data:

Table 8-1

		GPA	
		< 3.0	≥ 3.0
Amount of Sleep	< 8 hours	25	25
	> 8 hours	25	25

Since this was a cohort study, we calculate the risk ratio (RR):

$$RR = \frac{\frac{25}{50}}{\frac{25}{50}} = 1.0$$

There is no association between amount of sleep and subsequent GPA. Using the template sentence, this can be stated:

> Students who averaged fewer than 8 hours of sleep per night were 1.0 times as likely to end the term with a GPA below 3.0, compared to students who got at least 8 hours per night.

This is a risk ratio from a cohort study, so we need to include the time frame—which I did by saying "to end the term". Just as for confounding, we refer to this as the *unadjusted* or *crude* RR.

However, from talking to students, we wonder whether or not gender might be an important covariable. As with confounding, we would conduct a stratified analysis to check for effect modification. Again, we draw 2 × 2 tables with the same exposure (sleep) and outcome (GPA) but draw separate tables for men and women (gender is the covariable). We do this by looking back at the raw data and figuring out how many of the 25 people in the A (E+, D+) cell above were men and how many were women. Let's assume that of the 25 people who reported <8 hours and had a GPA < 3.0, 11 were men and 14 were women. We then similarly divide participants from the B, C, and D cells, and make stratum-specific 2 x 2 tables:

Table 8-2

Men		**GPA**	
		< 3.0	≥ 3.0
Amount of Sleep	< 8 hours	11	14
	8+ hours	17	9

Table 8-3

Women		GPA	
		< 3.0	≥ 3.0
Amount of Sleep	< 8 hours	14	11
	8+ hours	8	16

Effect Modification Example: Sleep and GPA, with gender as EM

Using data from the above 2×2 tables, the stratum-specific RRs are as follows:

$$RR_{men} = \frac{\left(\frac{11}{25}\right)}{\left(\frac{17}{26}\right)} = 0.68$$

$$RR_{women} = \frac{\left(\frac{14}{25}\right)}{\left(\frac{8}{24}\right)} = 1.7$$

Interpretations:

Among male students, those who slept fewer than 8 hours per night had 0.68 times the risk of having a GPA <3.0 at the end of the term, compared to those who reported 8 or more hours.

Among female students, those who slept fewer than 8 hours per night had 1.7 times the risk of having a GPA <3.0 at the end of the term, compared to those who reported 8 or more hours.

Sleeping fewer than 8 hours is associated—in these hypothetical data—with a *higher* GPA among male students (the "outcome" is low GPA, so an RR less than 1 indicates that exposed individuals are *less* likely to have a low GPA) but with a *lower* GPA among female students.

Gender in this case is acting as an effect modifier: the association between sleep and GPA varies according to strata of the covariable. You can spot effect modification when doing stratified analysis given the following:

- The stratum-specific measures of association are different than each other
- The crude falls in between them

If you have effect modification, the next step is to report the stratum-specific measures. We do not calculate an adjusted measure (it would be near 1.0, similar to the crude); the interesting thing here is that men and women react to sleep differently. Effect modification is something we want to highlight in our results, not something to be adjusted away.

How Different is Different?

Unlike for confounding, where a 10% change from crude to adjusted is an accepted definition for confounding, there exists no such standardized definition for how different the stratum-specific measures must be to call something an effect modifier. The threshold should probably be higher than the one necessary to declare something a confounder, because once you declare something an effect modifier, you are subsequently obligated to report results separately for each level of the covariable—something that cuts your **power** in at least half. Thus, in epidemiology, we rarely see evidence of effect modification reported in the literature. Long story short, "different" enough for effect modification is "unequivocally different."

When reading articles, effect modification will sometimes be called interaction, or the authors might just say that they are reporting stratified analyses. Any of these 3 phrases is a clue that there is a variable acting as an effect modifier.

Effect Modification Example II

Following the housing bubble-driven recession of 2008 (this is the exposure), the US economy lost a lot of jobs. Here is a graph showing the number of people who were working (the outcome) before, during, and after the recession. Results are being presented stratified by gender (a covariable), meaning the analyst suspected that gender might be acting as an effect modifier. Indeed, the results are slightly different: men (in blue) lost a greater proportion of jobs, and as of 2014 had not yet recovered to pre-recession levels, whereas women (in red) lost fewer jobs and by 2014 had fully recovered.

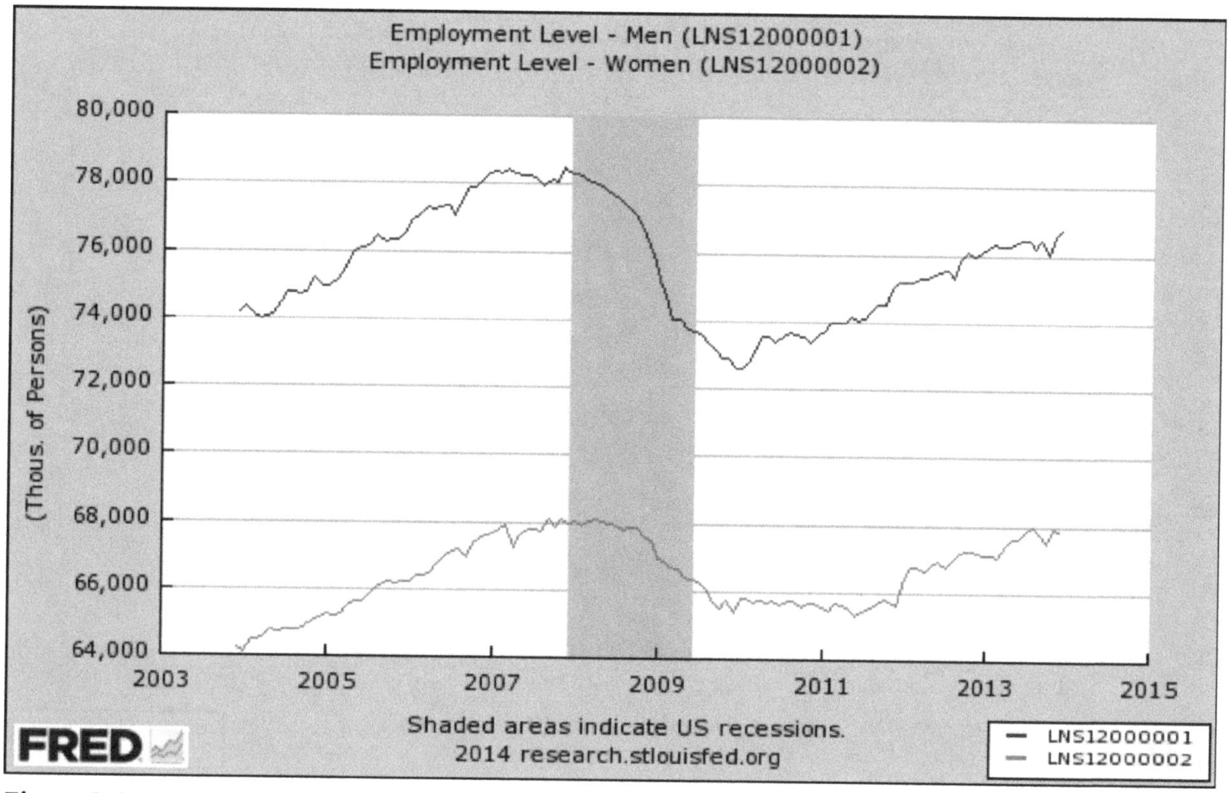

Figure 8-1
Source: *https://fred.stlouisfed.org/graph/fredgraph.png?g=qUs*

What if we also stratify by age? First, here is a graph showing how the recession affected jobs for people ages 55 and older:

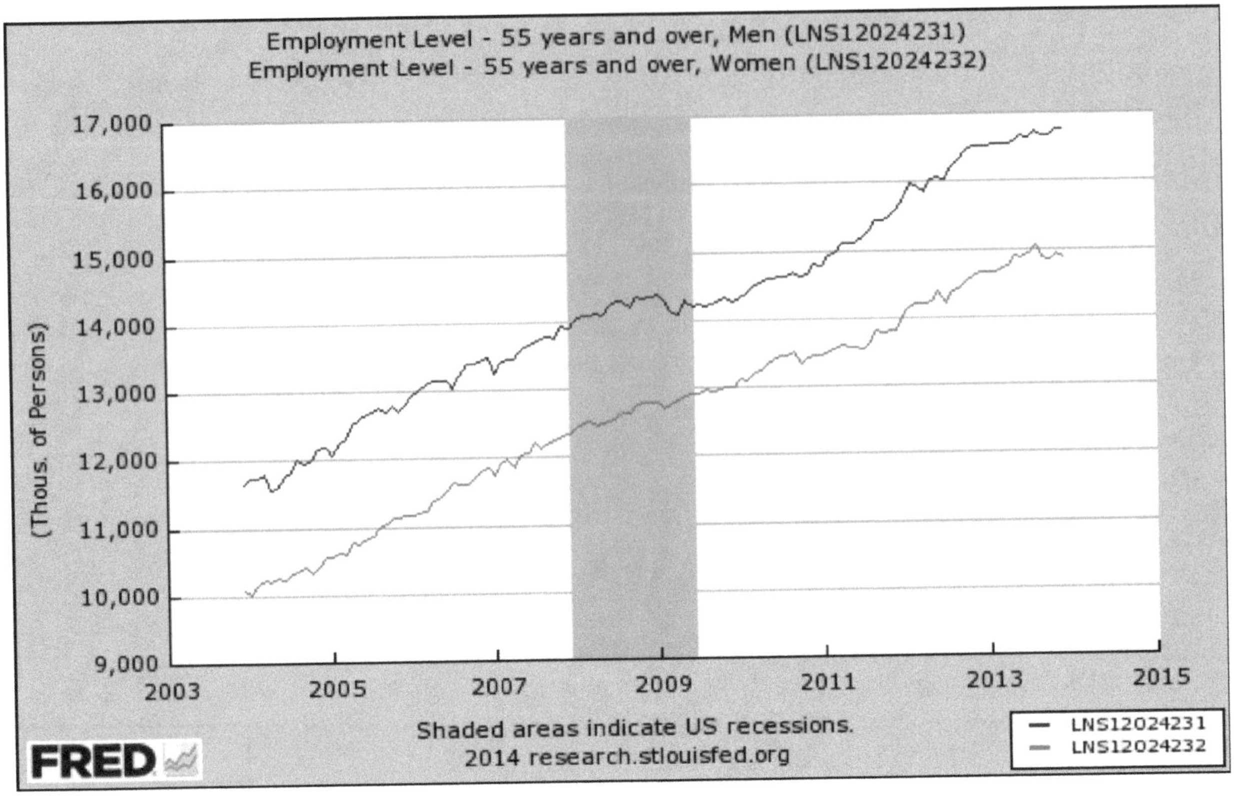

***Figure* 8-2**
Source: *https://fred.stlouisfed.org/graph/fredgraph.png?g=qUt*

The recession did not affect older working Americans at all. Nor are we seeing effect modification by gender—the 2 lines are nearly parallel.

What about young adults?

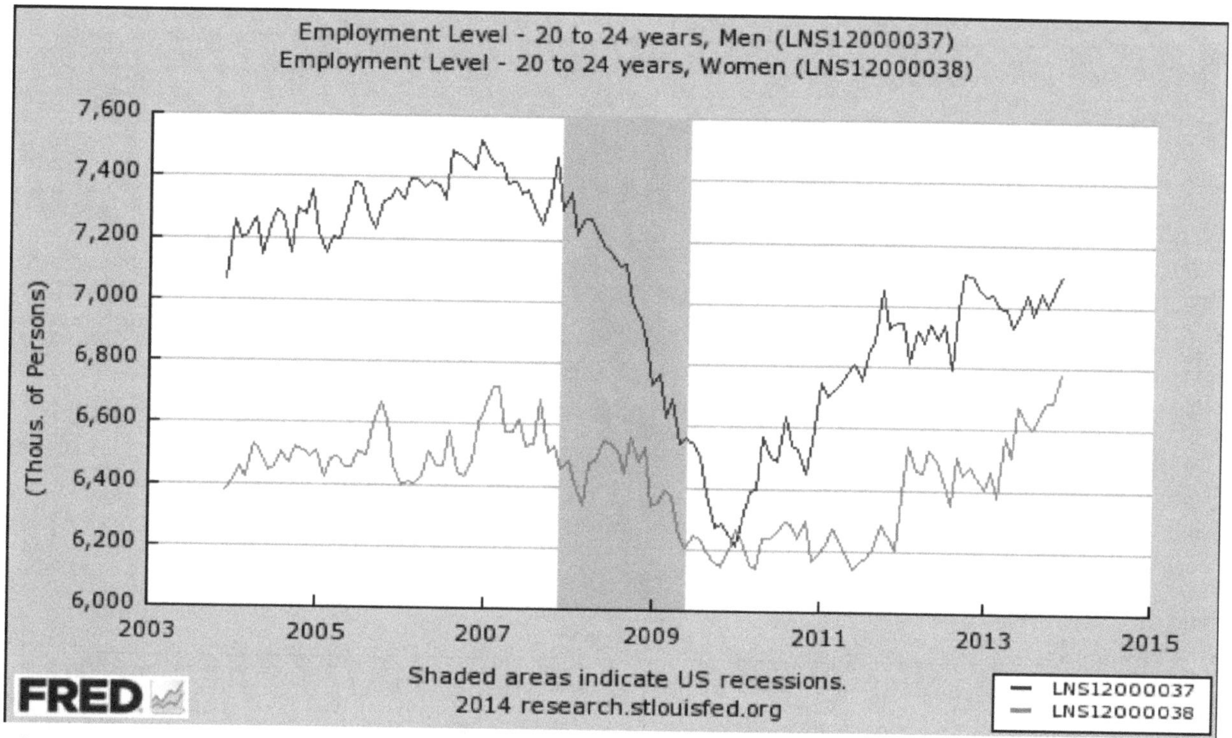

Figure 8-3
Source: *https://fred.stlouisfed.org/graph/fredgraph.png?g=qUv*

Here we have major effect modification by gender—young men lost a huge proportion of available jobs and had not recovered fully as of 2014. This is not surprising, as the recession was caused largely by the housing bubble, and construction workers are mostly young men. By contrast, young women lost a small proportion of jobs and quickly recovered to better-than-prerecession levels.

Finally, we look at jobs for 25 to 54-year-olds:

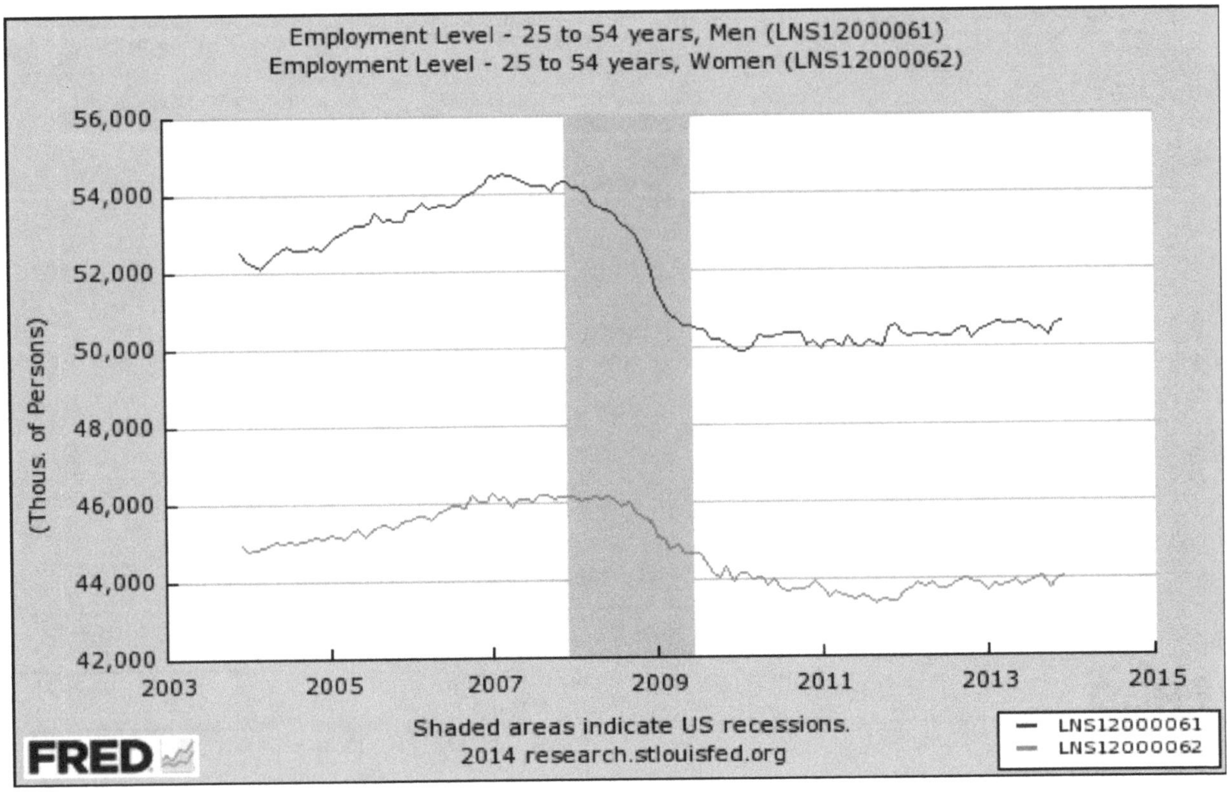

Figure 8-4
Source: *https://fred.stlouisfed.org/graph/?id=*LNS12000061

Here we see a very bleak picture. In this age group, jobs were lost—more for men than women—and as of 2014 had not recovered at all.

Thus when examining the job market's response to the 2008 recession, we see substantial effect modification by age (jobs recovery varied drastically by age) and, within some age categories, also some evidence of effect modification by gender. The effects of the recession on jobs were different for people of different ages and genders.

This is important because the policy implications would be very different. Imagine you were working as part of the federal government and trying to design an economic stimulus or recovery package. If the only data you had came from the first graph, without the age breakdowns, the potential policy solutions would be very different than if you also had access to the stratified-by-age analysis.

Differences between Confounding and Effect Modification

With confounding, you're initially getting the wrong answer because the confounder is not distributed evenly between your groups. This distorts the measure of association that you calculate (remember: having bigger feet is associated with reading speed only because of confounding by grade level). So instead you need to recalculate the measure of association, this time adjusting for the confounder.

With effect modification, you're also initially getting the wrong answer, but this time it's because your sample contains at least 2 subgroups in which the exposure/disease association is different. In this case, you need to permanently separate those subgroups and report results (which may or may not be confounded by still other covariables) separately for each stratum: in this case, men who sleep less have higher GPAs than men who sleep more, but at the same time, women who sleep more have higher GPAs than women who sleep less.

Here is a summary table denoting the process for dealing with potential confounders and effect modifiers. Much of the process is the same regardless of which type of covariable you have (in all cases, you must measure the covariable during your study, and measure it well!). Areas of difference are shown in **red**.

Table 8-4

	Confounding	Effect Modification
Before Planning a Study	Think about what variables might act as confounders based on what you know about the exposure/disease process under study.	Think about what variables might act as effect modifiers based on what you know about the exposure/disease process under study.
During a Study	Collect data about any potential covariables—stratified/adjusted analyses cannot be conducted without data on the covariable!	Collect data about any potential covariables—stratified/adjusted analyses cannot be conducted without data on the covariable!
Analysis: Step 1	Calculate the crude measure of association (ignoring the covariable).	Calculate the crude measure of association (ignoring the covariable).
Analysis: Step 2	Calculate stratum-specific measures of association, such that each level of the covariable has its own 2 x 2 table.	Calculate stratum-specific measures of association, such that each level of the covariable has its own 2 x 2 table.
Analysis: Step 3	**If the stratum-specific measures are similar to each other, and at least 10% different than the crude (which does not fall between them), then the covariable is a confounder.**	**If the stratum-specific measures are different than each other, and the crude lies between them, then the covariable is an effect modifier.**
Writing Results	**Report an adjusted measure of association that controls for the confounder.**	**Report the stratum-specific measures of association.**

Example III

Imagine that you do a cross-sectional study of physical activity and dementia in elderly people, and you calculate an unadjusted odds ratio (OR) of 2.0. You think that marital status might be an important covariable, so you stratify by "currently married" versus "not currently married" (which includes never married, divorced, and widowed). The OR among currently married people is 3.1, and among not currently married people the OR is 3.24. In this case, marital status is acting as a confounder, and we would report the adjusted OR (which would be 3.18 or so).

Example IV

Imagine that you do a randomized trial of a Mediterranean diet to prevent preterm birth in pregnant women. You do the trial and calculate an RR of 0.90. You think that perhaps **parity** might be an important covariable, so you conduct a stratified analysis. Among **nulliparas**, the RR is 0.60, and among **multiparas**, the RR is 1.15. These are different than each other, and the crude lies between them. In this case, parity is acting as an effect modifier, and so you would report the 2 stratum-specific RRs separately.

Example V

Imagine that you are doing a case-control study of melanoma and prior tanning bed use. The crude OR is 3.5, but perhaps gender is an important covariable. The stratified analysis yields an OR of 3.45 among men, and 3.56 among women. In this case, the covariable (gender) is neither a confounder nor an effect modifier. We say that it is not a confounder because (1) the crude lies between the 2 stratum-specific estimates, but also (2) the stratum-specific estimates are not more than 10% different than the crude. We say that it is not an effect modifier because, 3.45 and 3.56 are not that different—in both cases, there is a substantial effect (approximately 3.5 times as high). We would report the crude estimate of association, as it requires neither adjustment nor stratification to account for the effects of gender.

Can the Same Variable Act as Both a Confounder and an Effect Modifier?

Yes! Usually we see this when the covariable in question is a continuous variable, dichotomized for the purposes of checking for effect modification. For instance, if we think age might be an effect modifier, we might divide our sample into "old" and "young" for the stratified analysis—say, older than 50 versus 50 or younger. To the extent that 51-year-olds are not like 70-year-olds, we might miss some important nuances in the results, possibly because there exists in the data further effect modification with more categories (which would drop the power to almost nothing, were we to report separately on additional strata) or "residual" confounding as discussed in the previous chapter. Further details are beyond the scope of this book, but know that the same covariable can theoretically act as both a confounder and an effect modifier—but that one rarely sees this in practice.

Conclusion

Unlike confounding, whose effects we want to get rid of in our analysis, effect modification is an interesting finding in and of itself, and we report it. To check for effect modification, conduct a stratified analysis. If the stratum-specific measures of association are different than each other and the crude lies between them, then it's likely that the variable in question is acting as an effect modifier. Report the results separately for each stratum of the covariable.

One final, put-it-all-together table:

Table 8-5

If these are your ORs/RRs:				
Crude/Unadjusted	Stratum 1	Stratum 2	Then the covariable is…	And you would report…
2.0	1.0	3.2	an effect modifier	the 2 stratum-specific measures of association
2.0	3.5	3.6	a confounder	an adjusted measure
2.0	1.9	2.0	nothing interesting	the crude measure

9. Study Designs Revisited

Learning Objectives

After reading this chapter, you will be able to do the following:

1. Compare and contrast the strengths and limitations of cohort, case-control, cross-sectional, and randomized controlled trial studies
2. Describe ecologic studies and explain the ecologic fallacy
3. Describe the appropriate use of a systematic review and meta-analysis

Now that we have a firm understanding of potential threats to study validity, in this chapter we will revisit the 4 main epidemiologic study designs, focusing on strengths, weaknesses, and important details. I will also describe a few other study designs you may see, then end with a section on systematic reviews and meta-analyses, which are formal methods for synthesizing a body of literature on a given exposure/disease topic.

Cohorts

Recall from chapter 4 that a **cohort study** consists of drawing an at-risk (nondiseased) sample from the population, assessing levels of exposure, and then following the cohort over time and watching for incident disease:

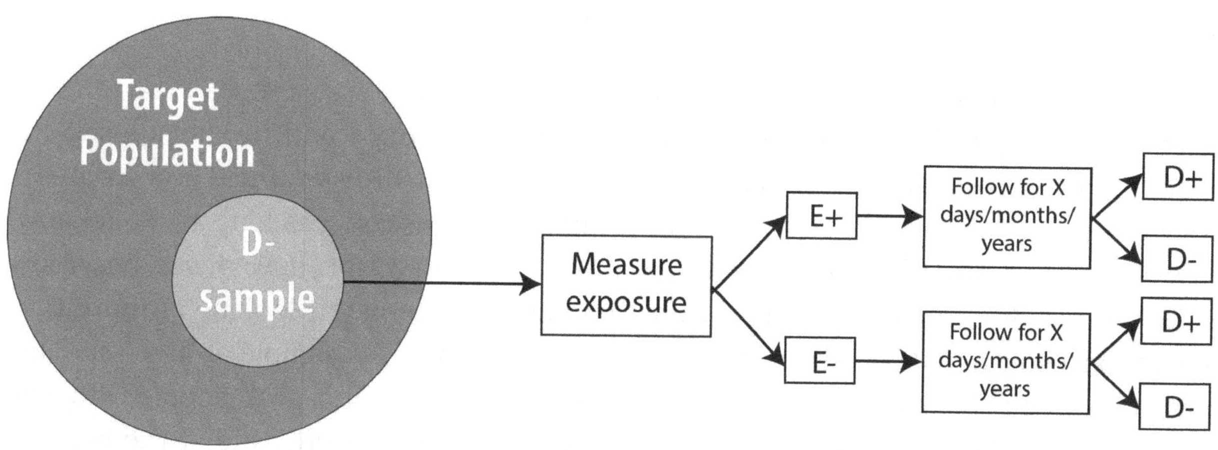

Figure 9-1

Cohort studies are a very strong study design, meaning that they are less prone to bias and temporality-related logical errors than some other designs. First, because we begin with a non-diseased sample, for which we immediately assess exposure status, we know that the exposure came first. Because of this, cohorts are unlikely to have misclassification of exposure differentially by disease status because the exposure is measured before disease status is known (misclassification of disease status differentially by exposure, however, is still a risk).

Cohort Temporality and Latent Periods

For diseases that have a long latent period—meaning that the biological onset of disease occurs long before the disease is detected and diagnosed—it is possible that some of our "nondiseased" sample are actually diseased but just have not been diagnosed yet. This could happen, for instance, for a cancer patient while their tumor is still too small to detect. When conducting studies on conditions with known or suspected long latent periods, epidemiologists will often exclude from the sample any participants in whom the disease is diagnosed during the first several months of follow-up, theorizing that those individuals were not truly disease-free at baseline.

Second, because cohort studies look for incident disease, they do not conflate the person's having the disease with how long they have had it, as prevalence studies do (see chapter 2 for a discussion of the mathematical relationship among incidence, prevalence, and duration of disease).

Third, they are the only study design that can be used to assess rare exposures. If the exposure is uncommon within the target population (say, 10% or fewer people can be expected to be exposed), then cohort studies can deliberately sample exposed individuals to ensure sufficient statistical **power** (the smallest cell in the 2 × 2 table drives the power) without needing an unreasonably large sample. For example, if we are concerned about chemical exposures in a particular factory, we might enroll exposed workers from that factory as well as a unexposed group of workers from a different factory (checking first, of course, to make sure that the second factory is truly exposure-free) and follow both groups, looking for incident disease.

Which leads nicely to the fourth strength: multiple outcomes can be assessed in the same cohort. In our factory example above, the exposed workers from Factory 1 and the unexposed workers from Factory 2 can be followed for any reasonably common disease. (Just how common is a judgment call—we could also watch for and track uncommon diseases, as long as we acknowledge that those analyses would be underpowered.) We could look for new-onset heart disease, leukemia, fibromyalgia, diabetes, death, or anything else of interest. If looking at more than one outcome, then we also must measure all outcomes of interest in the sample at baseline. Then, for analyses of each specific outcome, we merely eliminate from the cohort the people who were not at risk of that outcome. For instance, if Person A joins our factory study, and at the beginning of the study they already have **hypertension** but do not have melanoma, then we would not include that person in analyses where hypertension is the disease outcome. Their data could, however, be included in analyses where melanoma is the disease outcome, because they were at risk of melanoma at baseline.

Cohort studies can also be used to study multiple exposures, as long as these exposures are all common enough that we would not need to deliberately sample on exposure status. To do this, we would just grab a sample from the target population, and assess a multitude of exposures. If we want to also assess more than one outcome, then we need to measure all disease states of interest at baseline so that eventual analyses can be restricted to the population at risk, as discussed above. This ability to look at multiple outcomes—and potentially also multiple exposures—adds efficiency to cohort studies, as we can essentially conduct numerous studies all at once.

The Framingham Heart Study is a classic example of a cohort study that assessed multiple exposures and multiple outcomes. This study, a collaboration between the US National Heart, Lung, and Blood Institute (a division of the National Institutes of Health) and Boston University, began in 1948 by enrolling just over 5,000 adults living in Framingham, Massachusetts. Investigators measured numerous exposures and outcomes, then repeated the measurements every few years. As the cohort aged, their spouses, children, children's spouses, and grandchildren have been enrolled. The Framingham study is responsible for much of our knowledge about heart disease, stroke, and related disorders, as well as of the intergenerational effects of some lifestyle habits. More information and a list of additional publications (more than 3,500 studies have been published using Framingham data) can be found here.

Cohort studies also have downsides. They cannot be used to study rare diseases because the cohort would need to be too large to be practical. For example, phenylketonuria is a genetic metabolic disorder affecting about 1 in 10,000 infants born in the US.[1] To get even 100 affected individuals, then, we would need to enroll one million pregnant women in our study—a number that is neither practical nor feasible.

Furthermore, prospective cohort studies are costly. Following people over time takes a fair bit of effort, which means that study personnel costs are high. Because of this, cohort studies cannot be used to study diseases with decades-long **induction** or **latent periods**.

For example, it would be difficult to conduct a cohort study looking at whether adolescent dairy product consumption is associated with osteoporosis in 80-year-old women, because following current teenagers for 60 years or more would be extremely difficult. Along similar lines, selection biases related to lack of follow-up can be severe in studies with long durations: the longer we try to follow people, the more likely it is that they move, change phone numbers/email addresses, or get tired of filling out a survey every year and just stop participating. More troubling would be if people who start to feel ill are the ones who quit answering inquiries from the study team. What if these people were feeling ill because they were about to be diagnosed with the outcome under study? Despite this difficulty, a few long cohort studies such as Framingham exist and have yielded rich datasets and much knowledge about human health.

Randomized Controlled Trials

Recall from chapter 4 that an RCT is conceptually just like a cohort, with one difference: the investigator determines exposure status.

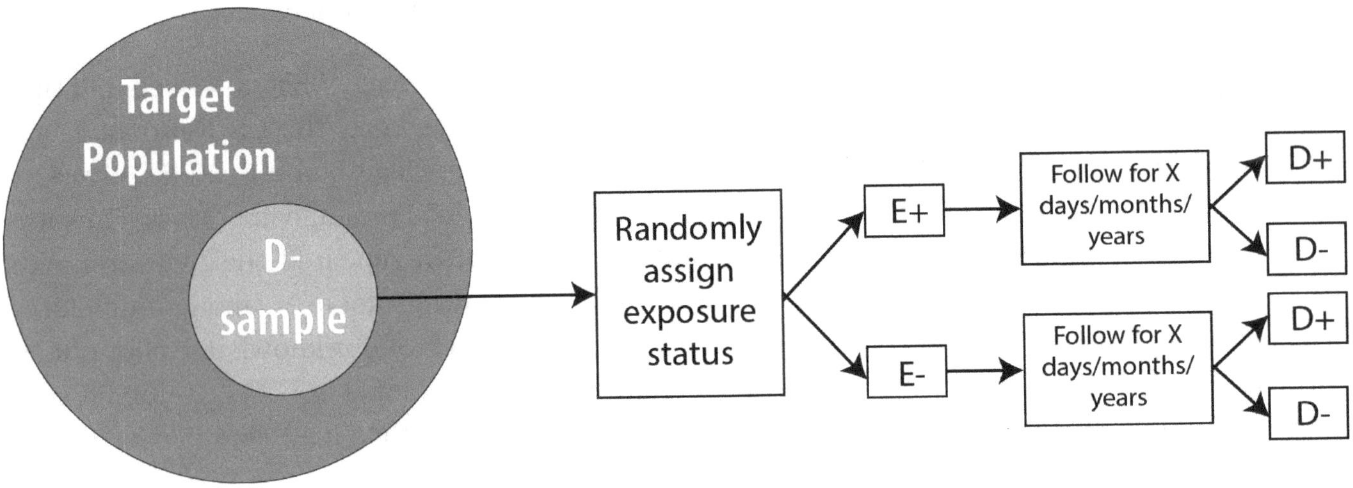

Figure 9-2

Thus all of the strengths and weaknesses of cohort studies apply also to RCTs, with one exception: to study multiple exposures, one would need to re-randomize for each exposure. A few studies have successfully done this (the Women's Health Initiative, for instance, randomized women to both hormone replacement therapy or placebo, and also, separately, to calcium supplements or placebo), but practically speaking RCTs are usually limited to one exposure.

One additional strength of a randomized trial (which does not apply to cohort studies) is that if the study is large enough (at least several hundred participants) and exposure allocation is truly random (i.e., not "every other person" or some other predictable scheme), then there will be no confounding. One can control, statistically, for measured confounders in a cohort study (see chapter 7), but what about any unknown and/or unmeasured confounders? The key feature of randomization is that it accounts for *all* confounders: known, unknown, measured, and unmeasured.

Recall from chapter 7 that for a variable to act as a confounder, it must satisfy these conditions:

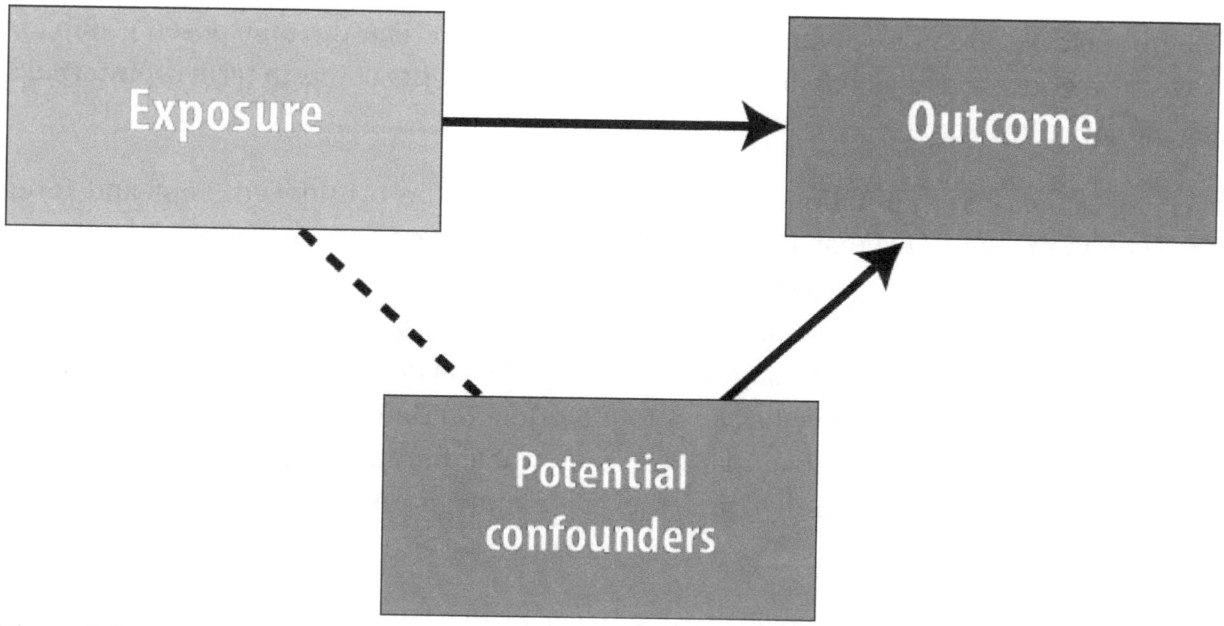

Figure 9-3

The variable must cause the outcome, be statistically associated with the exposure, and not be on the causal pathway (so the exposure does not cause the confounder). By randomly assigning the exposure, we have ensured that no variables exist that are associated with the exposure.

The picture now looks like this:

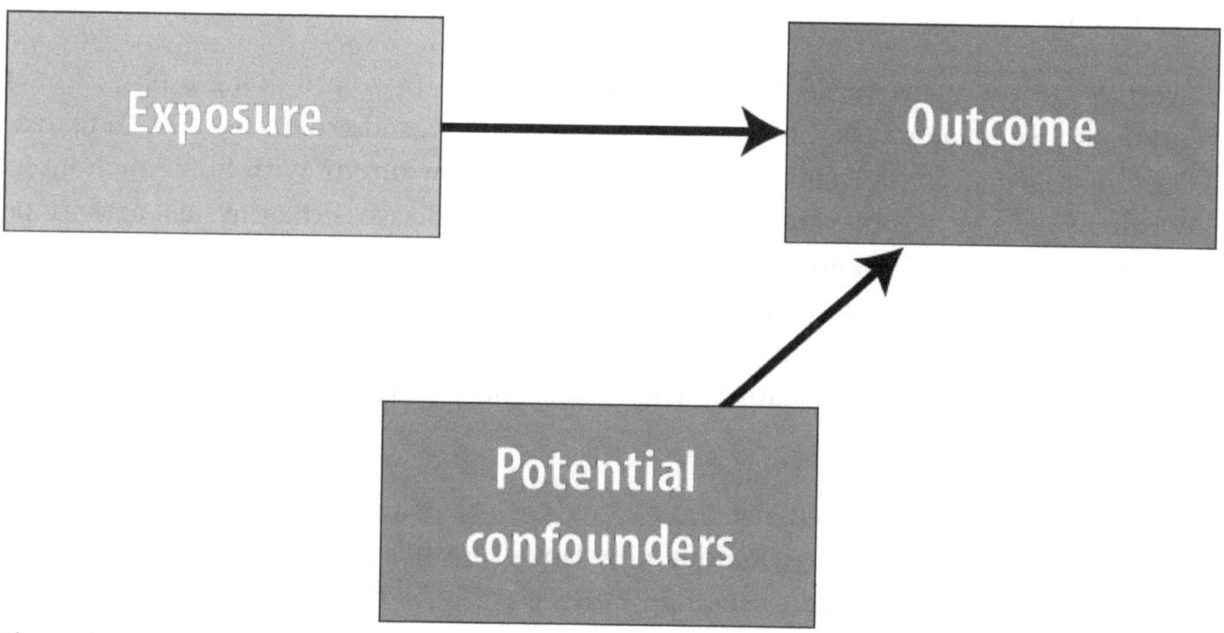

Figure 9-4

Because no variables are more common in the exposed group than the unexposed group (or vice versa), we have gotten rid of all possible confounding. *The benefits of this in terms of* ***internal study validity*** *cannot be overstated.*

However, RCTs also have limitations, and these should not be overlooked. First and foremost, they are even more expensive than cohort studies. Second, there are often ethical considerations rendering the randomized trial design unusable. For instance, at this point, we could not ethically justify randomizing people to a smoking exposure (because its harms are so well-documented, we cannot ask people to begin smoking for our study). We also cannot randomize where people live, but certainly where people live has a profound effect on their health.[ii] Observational studies of these exposures, on the other hand, are ethically viable because people have already chosen whether to smoke and where to live, and the epidemiologist merely measures these existing exposures.

Third, RCTs often have generalizability issues because the kinds of people who are willing to participate in a study where they (the participant) do not get to choose which study group to be in are not a random subset of the overall population. For instance, if the only people who have time to participate in our physical activity intervention are people who are retired, then can we generalize to the (presumably younger) population who are still working? Perhaps—but perhaps not. Investigators conducting RCTs also sometimes overly restrict the inclusion criteria to the extent that results are not generalizable to the overall population. For instance, a well-known trial of blood pressure control in older adults excluded those with diabetes, cancer, and a host of other comorbidities.[iii] Given that most older people have at least one of these chronic diseases, to whom can we really apply the results?

Lastly, we have to precisely specify the exposure in an RCT. If we are doing a physical activity intervention, are we going to ask those randomized to the exposed group to walk? To take a yoga class? Do supervised strength training? If so, how much? How often? With how much intensity? For how many weeks or months? In a cohort study, we would assess the physical activity people are doing anyway, and there would be a huge variety of responses, which we could then categorize in any number of ways. With a randomized trial, we have to decide on all of the details. If we are wrong, or if we apply the intervention at the wrong time in the disease process, it could seem like there is no exposure/disease association, when really there is and our exposure was slightly off somehow.

Randomized trials are often called the "gold standard" of epidemiologic and clinical research because of their ability to minimize confounding. However, their drawbacks are substantial, and well-conducted observational studies should not necessarily be discounted merely because they are not RCTs. Nonetheless, RCTs play an enormous role particularly in medicine, as the Food and Drug Administration (FDA) requires multiple RCTs prior to approving new drugs and medical devices. Because of the FDA's strict requirements, protocols for randomized trials must be registered (at clinicaltrials.gov) prior to the start of any data collection.

Outside of pharmaceutical research and development, RCTs, because of their methodologic strengths, have the potential to change practice when evidence from new, large, well-designed studies emerge. For example, in 2005 Dr. Paul Ridker and colleagues absolutely changed the way physicians thought about heart disease **prophylaxis** in women.[iv] Prior to publication of this large (20,000 women in each group) trial, we assumed that, like men, older women should take a baby aspirin every other day to prevent heart attacks. However, the Ridker trial showed that aspirin acts differently in women (gender is an effect modifier!), and the aspirin-a-day-prevents-heart-attacks regimen will not work for most women.

Case-Control Studies

A case-control study is a retrospective design wherein we begin by finding a group of cases (people who have the disease under study) and a comparable group of controls (people who do not have the disease):

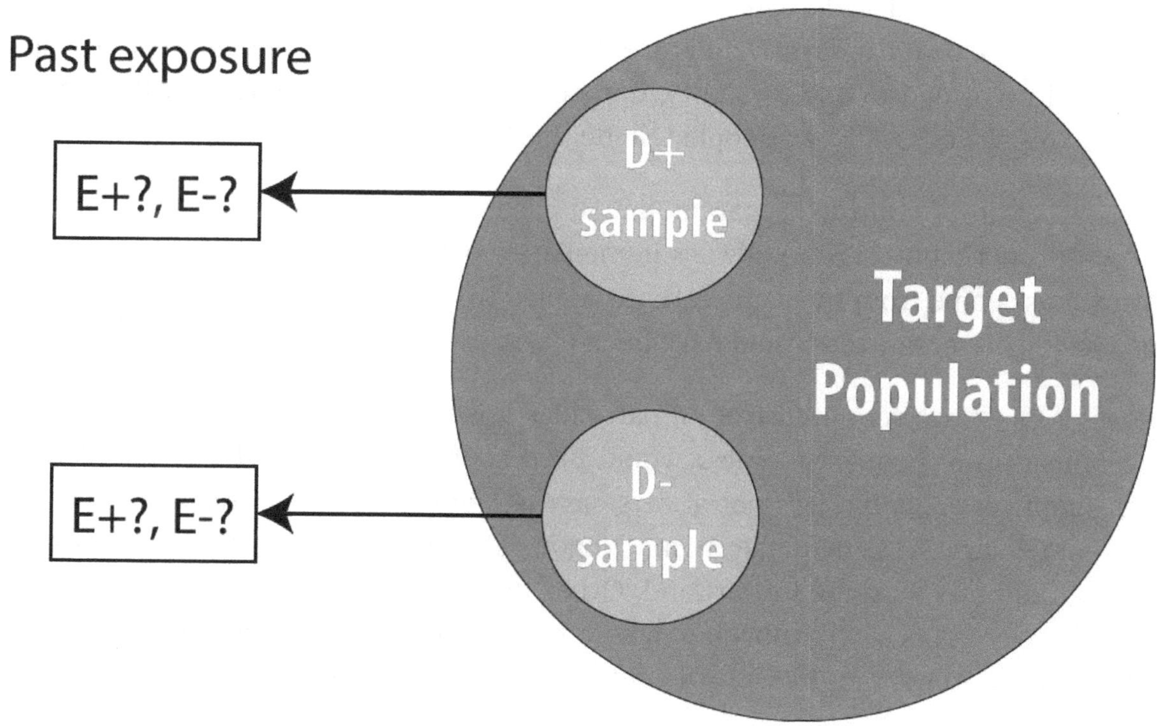

Figure 9-5

A common mistake made by beginning epidemiology students is to state that "cases are people with the disease, who are exposed." This is incorrect. Cases are people with the disease, and to avoid differential misclassification, it is important that both cases and controls be recruited without regard for exposure status. Once we have identified all cases and controls, then we assess which people were exposed.

Because they do not require following people over time, case-control studies are much cheaper to conduct than cohorts or randomized trials. They also provide an efficient way to study rare diseases and diseases with long induction and/or latent periods. Case-control studies can assess multiple exposures, though they are limited to one outcome by definition.

Case-control studies assess exposure in the past. Occasionally, these past exposure data come from existing records (e.g., medical records for a person's blood pressure history), but usually we rely on questionnaires. Case-control studies are thus subject to recall bias, more so than prospective designs. Epidemiologists conducting case-control studies need to be particularly wary of differential recall by case status. It is plausible that people with a given condition will have spent time thinking about what might have caused it and thus be able to report past exposures with greater detail than members of the control group. Regardless of case status, the questions asked must be possible for people to answer. No one can say with certainty exactly what they ate on a particular day a decade ago; however, most people can probably recall what kinds of foods they usually ate on most days. Details are thus sacrificed in favor of bigger-picture accuracy (which may still be of questionable validity, depending on people's memories). Remember from chapters 5 and 6: ask yourself, "Can people tell me this? Will people tell me this?"

The proper selection of controls is paramount in case-control studies, but unfortunately, who constitutes a "proper" control is not always immediately obvious. To avoid selection biases, cases and controls must come from the same target population—that is, if controls had been sick with the disease in question, they too would have been cases.

For instance, if cases are recruited from a particular hospital, then controls should be sampled from the population of people who also would have sought care at that hospital if necessary. This seems simple enough, but it is not always easy to translate into practice. If we are studying traumatic brain injury (TBI) in children in Oregon, a good place to find cases would be at Doernbecher Children's Hospital in Portland. Other hospitals throughout the Pacific Northwest send kids with severe TBIs to Doernbecher, where a myriad of pediatric specialists are available to care for them; this hospital thus has a sufficient number of cases for our study.

Where would we get controls? One possibility would be to take as controls other children who are patients at Doernbecher, for a condition other than TBI. This satisfies the criterion that controls would also get care at this hospital, because they *are* getting care at this hospital. However, to the extent that kids receiving care for other conditions might *also* have unusual exposure histories, this could lead to biased estimates of association. Another option would be to designate as controls children who are not sick, sampled perhaps from a Portland neighborhood or two. However, this would also lead to selection bias, because Doernbecher is a referral hospital, receiving as patients children from a several-hundred-mile radius, not just children who live in Portland. If kids who live out in more rural areas are different than those who live in the city, we would have biased estimates of association.

The bottom line is that there is no perfect way to recruit controls, and epidemiologists love to poke holes in other people's control groups for case-control studies[v,vi] (this is considered good sport at epidemiology conferences). One way to reduce bias from the control group is to recruit multiple control groups—perhaps one hospital-based and one community-based. If the results are not substantially different, then any selection biases that are operating are perhaps not overly influencing the results.

For long-lasting chronic diseases, the issue of disease duration again comes into play. To avoid temporality issues, we must know at a minimum the date of diagnosis and ensure that we are assessing exposures that happened well before that date. For conditions for which the induction and latent periods are unknown, investigators will sometimes conduct a case-control study that recruits incident cases of disease over a period of several months. Thus as soon as cases are recruited, we can ask about past exposures with the confidence that at least the case diagnosis occurred after those exposures. While a long latent period might still be an issue, one way around this would be to ask about exposures over multiple time periods—say, 0–5 years ago, 6–10 years ago, 11–15 years ago, and so on—and compare results across these windows.

Despite these difficulties, case-control studies have made substantial contributions to our knowledge about health over the years. The surgeon general's 1964 report *Smoking and Health*[vii], for instance, was based on literature that stemmed from a case-control study conducted by Richard Doll and Austin Bradford Hill.[viii]

Cross-Sectional Studies

Recall from chapter 4 that in a cross-sectional study, we draw a single sample from the target population and assess current exposure and disease status on everyone:

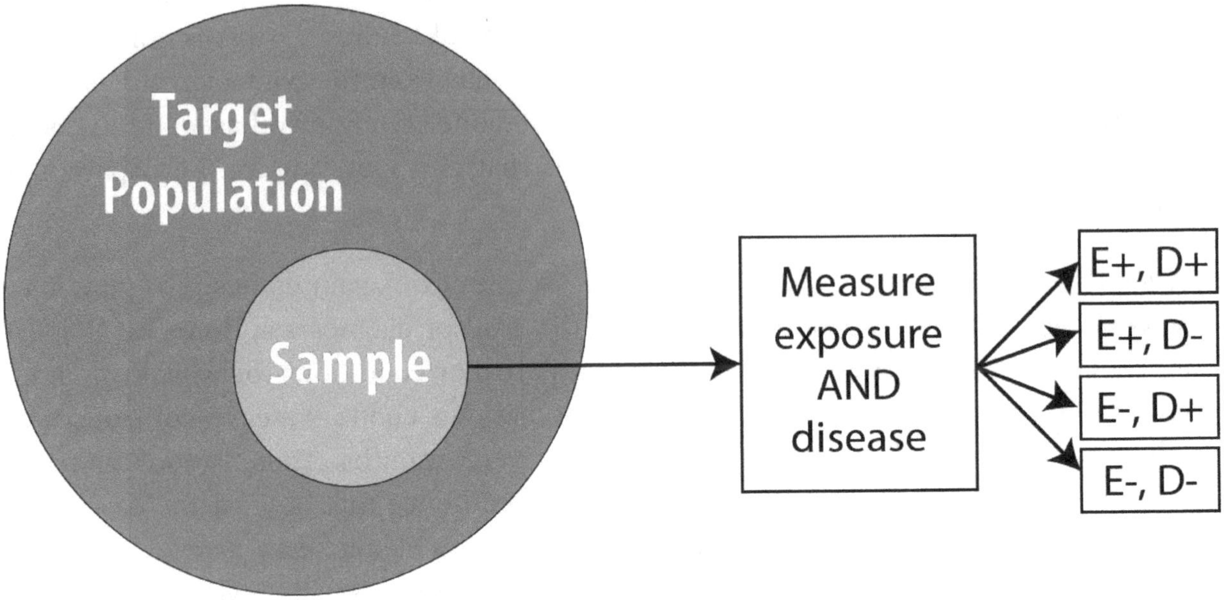

Figure 9-6

The main strength of cross-sectional studies is that they are the fastest and cheapest studies to conduct. They are thus used for many surveillance activities—the National Health and Nutrition Examination Survey (NHANES), Pregnancy Risk Assessment Monitoring System (PRAMS), and Behavioral Risk Factor Surveillance System (BRFSS) are all cross-sectional studies that are repeated with a new sample each year (see chapter 3)—and in other situations where resources may be limited and/or immediate answers are required.

Cross-sectional studies are limited by the fact that we sample for neither exposure nor disease and that we instead "get what we get" when drawing our sample from the population. They thus cannot be used for either rare exposures or rare diseases.

Another limitation is that we have no data on temporality: we do not know whether the exposure or the disease came first because we are measuring the prevalence of both at the same point in time.

Cross-sectional studies along with surveillance (which looks only at measures of disease frequency, not at exposure/disease relationships) are thus limited to hypothesis generation activities. We cannot make (nonsurveillance) public health or clinical decisions based on evidence only from these studies.

Case Reports/Case Series

In the clinical literature, one often sees case reports. These are short blurbs reporting an interesting and unusual patient seen by a particular doctor or clinic. A case series is the same thing but describes more than one patient—usually only a few,[ix,x] but sometimes several hundred.[xi] Case reports and case series have little value for epidemiologists because they are not studies per se; they have no comparison groups. If a case series is published saying that 45% of patients in this series with disease Y also have disease Z, this is not useful information for an epidemiologist. How many patients who do *not* have disease Y also have disease Z? Without data on a comparable group of patients who do not have disease Y, there is nothing to be done with the 45% data point given in the case series.

That said, case reports and case series can be extremely useful for public health professionals. Because by definition they present data from unusual patients, they can often act as a kind of sentinel surveillance, drawing our attention to a new, emerging public health threat. For example, in 1941, a physician from Australia noticed an increase in a kind of birth defect affecting infant eyes. He published this as a case series,[xii] hypothesizing that maternal **rubella** infection was the cause. Other physicians from around the world chimed in that they, too, had seen a recent sudden increase in this birth defect in women whose pregnancies were complicated by rubella,[xiii,xiv,xv] leading to our current practice of checking for rubella antibodies in all pregnant women and vaccinating those without immunity. As another example, in the early 1980s, a set of case series published by the Centers for Disease Control and Prevention (CDC) in its *Morbidity and Mortality Weekly Report* drew our attention to unusual kinds of cancers and opportunistic infections occurring in otherwise young, healthy populations—our first inkling of the HIV/AIDS epidemic.[x,xvi,xvii,xviii] More recently, in 2003, case reports detailing an unusual, deadly respiratory infection in people traveling to Hong Kong led to increased global public health and clinical awareness of this unusual set of symptoms, allowing immediate quarantine of affected individuals who had traveled back to Toronto.[xix,xx,xxi] This quick action prevented SARS from becoming a global pandemic.

Ecologic Studies

Ecologic studies are those in which group-level data (usually geographic) are used to compare rates of disease and/or disease behaviors. For instance, this picture showing variation in seat belt use by state from chapter 1 is a kind of ecologic study:

Figure 9 7 Source: *https://www.cdc.gov/motorvehiclesafety/seatbelts/seatbelt_map.html*

By comparing rates of seat belt use across different states, we are comparing group-level data, not data from individuals. While useful, this kind of picture can lead to many errors in logic. For example, it assumes that everyone in a given state is exactly the same—obviously this is not true. While it is true that on average, people in Oregon wear their seat belt more often than people from Idaho, this does not mean that everyone in Oregon wears their seat belt more often than everyone in Idaho. We could easily find someone in Oregon who never wears their seat belt and someone in Idaho who always does.

The above logical error—ascribing group-level numbers to any one individual—is an example of the **ecologic fallacy**. This also comes into play when looking at both exposure and disease patterns using group-level data, as in this example, looking at per-capita rice consumption and maternal mortality in each country:

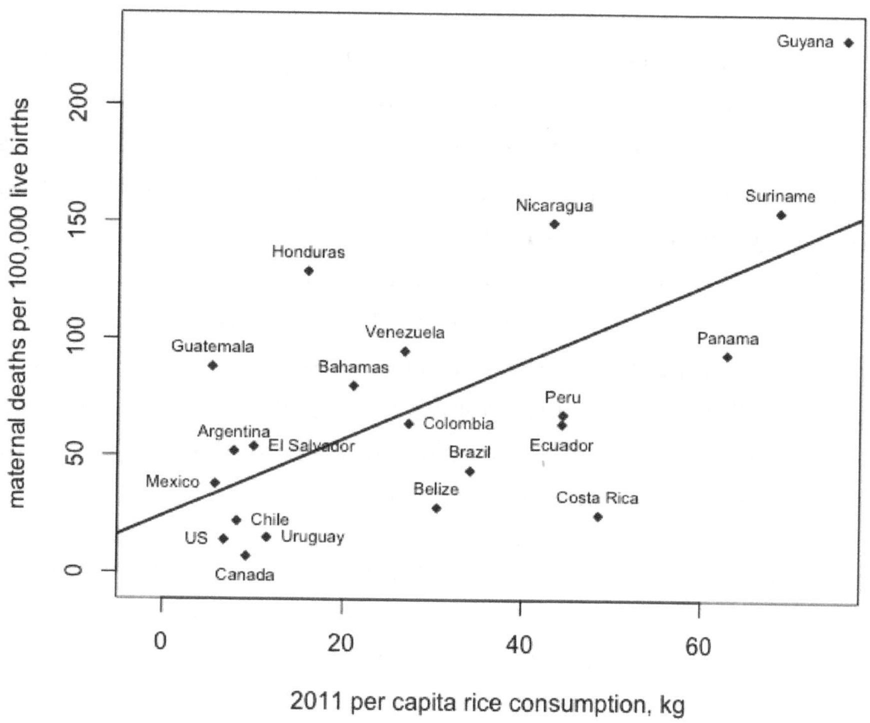

Figure 9-8. Created with data from here and here.

From looking at this graph, it appears that the more rice that is consumed by citizens of a particular country, the higher the maternal mortality rate. The ecologic fallacy here would stem from assuming that it is the rice consumers who are dying from complications related to pregnancy or birth, but we cannot know whether this is true using only group-level data.

With all of these problems, then, why conduct ecologic studies? Even more so than cross-sectional studies, they are quick and cheap. They also always use preexisting data—census estimates for per-county income; the amount of some product (such as rice) consumed by a given group of people (often tracked by sellers of that product); and recorded information on the prevalence of certain diseases (usually publicly available via the websites of health ministries for various countries or as by-country comparisons published by the World Health Organization). The use of ecologic studies is limited only to hypothesis generation, but they are so easy that they can be a good first step for a totally new research question.

Systematic Reviews and Meta-analyses

Because epidemiology relies on humans, it is more prone to both bias and confounding than other sciences. Does this render it useless? Absolutely not, though one must have a robust appreciation for the assumptions and limitations inherent in epidemiologic studies. One of these limitations is that barring exceptionally well-done, randomized controlled trials (as the Ridker trial,[iv] mentioned previously), we rarely change public health or clinical policy based on just one epidemiologic study. Rather, we do one study, then another, and then another, using better and better study designs until eventually there is a body of evidence on a topic that stems from different populations, uses different study designs, perhaps measures the exposure in slightly different ways, and so on. If all these studies tend to show the same general results (as did all the early studies on smoking and lung cancer), then we start to think that the association might be causal (see chapter 10 for more detail on this) and implement public health or clinical changes.

When results of existing studies on a topic are more mixed, there is a formal way of synthesizing their results across all of them, to arrive at "the" answer: meta-analysis (or systematic review—they differ slightly, as discussed below). The procedure for either of these is the same:

1. Determine the topic—precisely. Do we care about correlates of physical activity in kids generally, or only in PE class at school? Only at home? Everywhere? Is our focus all children or only grade-school kids? Only adolescents? There often is no right answer, but as with defining our target population (see chapter 1), this needs to be decided ahead of time.
2. Systematically search the literature for relevant papers. By systematically, I mean using and documenting specific search terms and placing documented limits (language, publication date, etc.) on the search results. The key is to make the search replicable by others. It is not acceptable to just include papers that authors are aware of without searching the literature for others—doing so results in a biased sample of all the papers that should have been included.
3. Narrow down the search results to only those directly addressing the topic as determined in Step 1.
4. For each of the studies to be included, abstract key data: the exposure definition and measurement methods, the outcome definition and measurement methods, how the sample was drawn, the target population, the main results, and so on.
5. Determine whether the papers are similar enough for meta-analysis (there are formal statistical procedures to test for this, which are beyond the scope of this book).[xxii(p287)]
    1. If they are, then researchers essentially combine all the data from all the included studies and generate an "overall" measure of association and 95% confidence interval.
    2. If they are not, then the authors will synthesize the studies in other meaningful ways,

comparing and contrasting their results, strengths, and weaknesses and arriving at an overall conclusion based on the existing literature. An overall measure of association is *not* calculated, but usually the authors are able to conclude that some exposure either is or is not associated with some outcome (and perhaps roughly the strength of that association).

6. Assess the likelihood of **publication bias** (again, there are formal statistical methods for this)[xxii(pp197-200)] and the degree to which that may or may not have affected the results.
7. Publish the results!

Ideally, at least 2 different investigators will conduct steps 2–4 completely independently of each other, checking in after completion of each step and resolving any discrepancies, usually by consensus.[xxiii] This provides a check against un- or subconscious bias on the part of the authors (remember: we're all human and therefore all biased). For systematic reviews and meta-analyses conducted after 2015 or so, the protocol for the review (search strategy, exact topic, etc.) should be registered prior to step 2 with a central registry, such as PROSPERO. This provides a check against bias—authors who deviate from their preregistered protocols should provide very good reasons for doing so, and such studies should be interpreted with extreme caution.

Results from meta-analyses are often presented as forest plots that plot each included study's main result (with the size of the square corresponding to sample size) and an overall estimate of association is indicated as a diamond at the bottom. Here is an example from a meta-analysis of chocolate consumption and systolic blood pressure (SBP, the top number in a blood pressure reading):

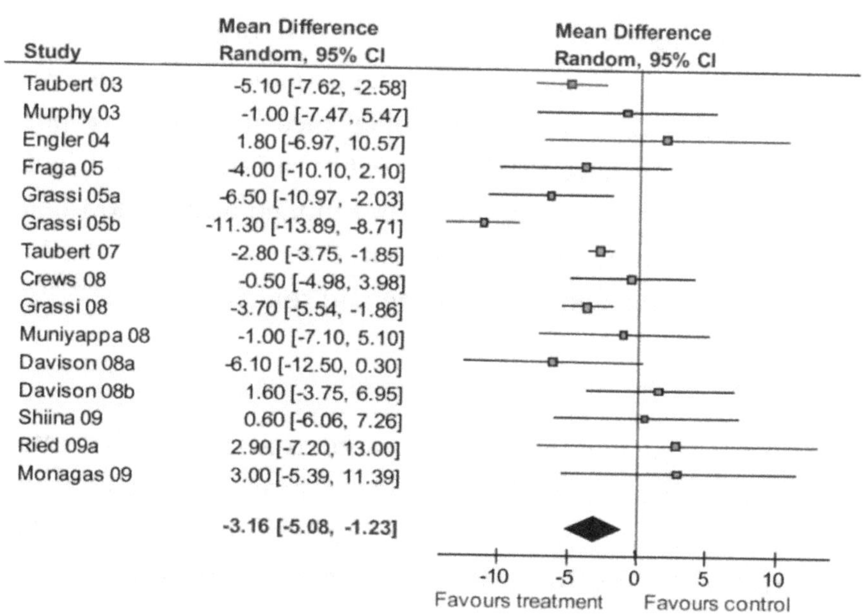

Figure 9-9. Adapted from Reid et.al., BMC Medicine 2010

You can see from this forest plot that the majority of studies showed a decrease in SBP for people who ate more chocolate, though not all studies found this. Some point estimates are quite close to 0.0 (which is the "null" value here, because we're looking at change in a single number, not a ratio), and 10 of the confidence intervals cross 0.0, indicating that they are not statistically significant. Nonetheless, 6 studies–the largest studies, since their confidence intervals are the narrowest–*are* statistically significant, and all of these in the direction of chocolate being beneficial. Indeed, the overall (or "pooled") change in SBP and 95% CI shown at the bottom (the black diamond) indicates a small (approximately 3 mm Hg) reduciton in SBP for chocolate consumers. Does this mean we should all start eating lots of chocolate? Not necessarily: a 3 mm Hg ("millimeters of mercury"–still the units in which we measure blood pressure, despite mercury not being involved for several decades now) drop in SBP is not *clinically significant*. A normal SBP is between 90 and 120, so a 3 mm Hg drop puts you at 87-117–likely not even a noticeable physiologic change.

As alluded to above, meta-analysis requires a certain similarity among the studies that will be pooled (e.g., they need to control for similar, if not identical, confounders). Often, this is not the case for a given body of literature—in which case, the authors will systematically examine all the evidence and do their best to come up with "an" answer, taking into consideration the quality of individual studies, the overall pattern of results, and so on. For example, in a systematic review of risk-reducing mastectomy (RRM), the prophylactic surgical removal of breasts in women who do not yet have breast cancer, but who have the BRCA-1 or BRCA-2 genes and thus are at very high risk (where BRRM refers to *bilateral* RRM—having both breasts removed). The authors described the overall results of this study as follows:

> Twenty-one BRRM studies looking at the incidence of breast cancer or disease-specific mortality, or both, reported reductions after BRRM, particularly for those women with BRCA1/2 mutations....Twenty studies assessed psychosocial measures; most reported high levels of satisfaction with the decision to have RRM but greater variation in satisfaction with cosmetic results. Worry over breast cancer was significantly reduced after BRRM when compared both to baseline worry levels and to the groups who opted for surveillance rather than BRRM, but there was diminished satisfaction with body image and sexual feelings.[xxvii(p2)]

The authors then concluded:

> While published observational studies demonstrated that BRRM was effective in reducing both the incidence of, and death from, breast cancer, more rigorous prospective studies are suggested. [Because of risks associated with this surgery] BRRM should be considered only among those at high risk of disease, for example, BRCA1/2 carriers.[xxvii(p2)]

No overall "pooled" estimate of the protective effect associated with RRM is provided, but the authors are nonetheless able to convey the overall state of the literature, including where the body of literature is lacking.

Systematic reviews and meta-analyses are excellent resources for learning about a topic. Realistically, no one has the time to keep up with the literature in anything other than a very narrow topic area, and even then it is really only a boon to researchers in that field to take note of new individual studies. For public health professionals and clinicians not routinely engaging in research, relying on systematic reviews and meta-analyses provides a much better overall picture that is potentially less prone to the biases found in individual studies. However, care must be taken to read *well-done* reviews. The title of the paper should include either *systematic review* or *meta-analysis*, and the methods should mirror those outlined above. Be wary of review papers that are not explicitly systematic—they are extremely prone to biases on the part of the authors and probably should be ignored.[1]

Conclusions

The figure below is a representation of the relative cost and internal validity of the study designs discussed in this chapter:

1. *Metasynthesis* is a legitimate technique for systematic reviewing qualitative literature. The papers to watch out for are the ones called "integrative review," "literature review," or just "review"—anything that is not "systematic review."

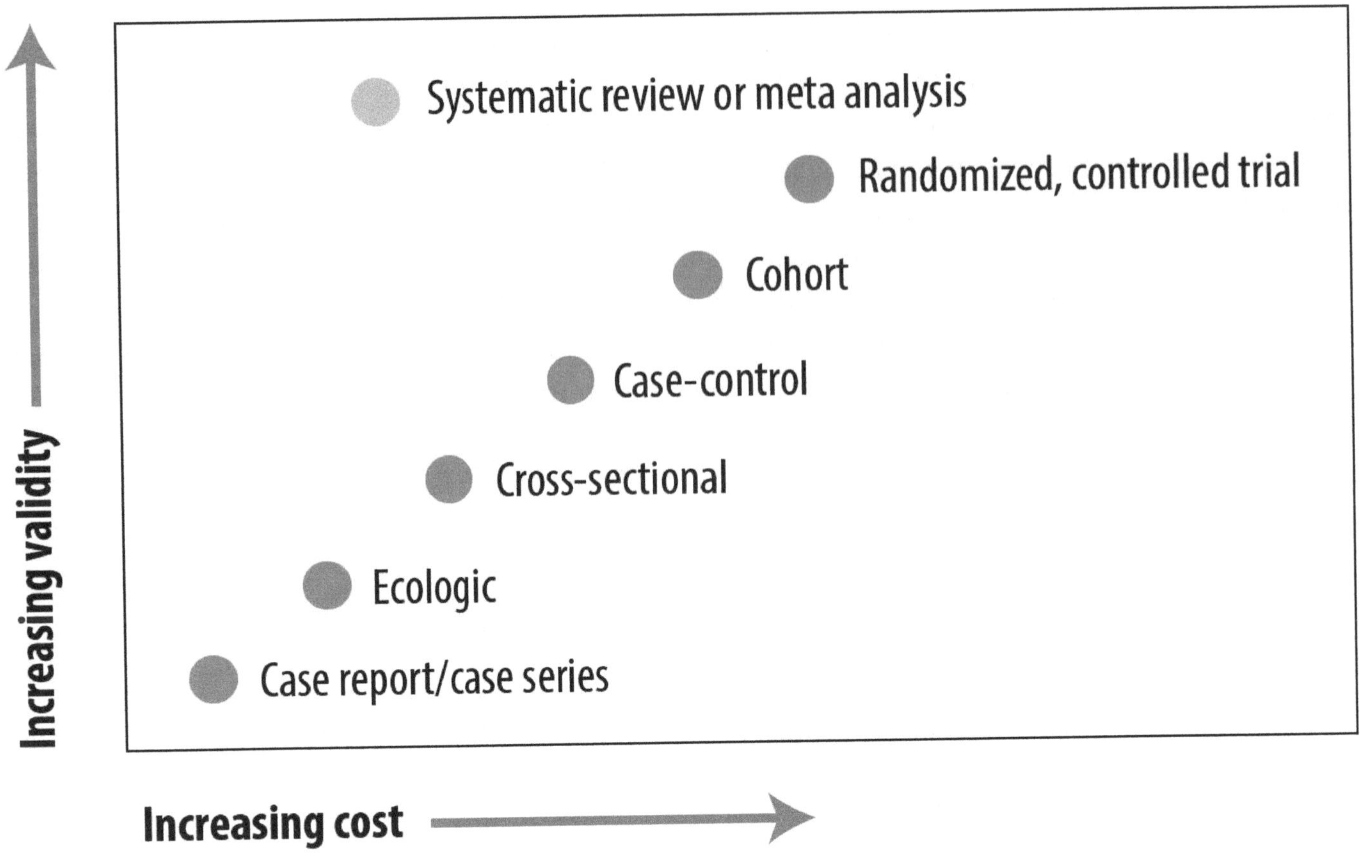

Figure 9-10

There are many types of epidemiologic studies, from reports of a single, unusual patient up to formal meta-analyses of dozens of other studies. The relative validity of these in terms of using their evidence to shape policy varies widely, but with the exception of review papers, the "better" studies are the more expensive and time-consuming ones. Review papers in and of themselves are not particularly expensive, but they cannot be done until numerous other studies have been published, so if you include those as indirect costs, they take a lot of time and money. The 4 main study types (cross-sectional, case-control, cohort, and RCT) each have strengths and weaknesses, and readers of the epidemiologic literature should be aware of these. There are occasions, independent of cost or validity considerations, when one design or another is preferred (e.g., case-control for rare diseases).

References

i. Williams RA, Mamotte CD, Burnett JR. Phenylketonuria: an inborn error of phenylalanine metabolism. *Clin Biochem Rev*. 2008;29(1):31-41.

ii. Could where you live influence how long you live? RWJF. https://www.rwjf.org/en/library/interactives/whereyouliveaffectshowlongyoulive.html. Accessed February 19, 2019.

iii. A randomized trial of intensive versus standard blood-pressure control. N *Engl J Med*. 2017;377(25):2506. doi:10.1056/NEJMx170008

iv. Ridker PM, Cook NR, Lee I-M, et al. A randomized trial of low-dose aspirin in the primary prevention of cardiovascular disease in women. N *Engl J Med*. 2005;352(13):1293-1304. doi:10.1056/NEJMoa050613. (↲ Return 1) (↲ Return 2)

v. Wacholder S, McLaughlin JK, Silverman DT, Mandel JS. Selection of controls in case-control studies, I: principles. *Am J Epidemiol*. 1992;135(9):1019-1028. (↲ Return)

vi. Wacholder S, Silverman DT, McLaughlin JK, Mandel JS. Selection of controls in case-control studies, II: types of controls. *Am J Epidemiol*. 1992;135(9):1029-1041. (↲ Return)

vii. Health CO on S and. Smoking and tobacco use: history of the Surgeon General's Report. Centers for Disease Control and Prevention. 2017. http://www.cdc.gov/tobacco/data_statistics/sgr/history/. Accessed October 30, 2018. (↲ Return)

viii. Doll R, Hill AB. Smoking and carcinoma of the lung. *Br Med J*. 1950;2(4682):739-748. (↲ Return)

ix. Bowden K, Kessler D, Pinette M, Wilson E. Underwater birth: missing the evidence or missing the point? *Pediatrics*. 2003;112(4):972-973.

x. Centers for Disease Control and Prevention (CDC). A cluster of Kaposi's sarcoma and Pneumocystis carinii pneumonia among homosexual male residents of Los Angeles and Orange counties, California. *MMWR Morb Mortal Wkly Rep*. 1982;31(23):305-307.

xi. Cheyney M, Bovbjerg M, Everson C, Gordon W, Hannibal D, Vedam S. Outcomes of care for 16,924 planned home births in the United States: the Midwives Alliance of North America statistics project, 2004 to 2009. *J Midwifery Womens Health*. 2014;59(1):17-27. (↲ Return)

xii. Gregg NM. Congenital cataract following German measles in the mother. *Trans Opthalmol Soc Aust*. 1941;3:35-46. (↲ Return)

xiii. Greenberg M, Pellitteri O, Barton J. Frequency of defects in infants whose mothers had rubella during pregnancy. *J Am Med Assoc*. 1957;165(6):675-678. (↲ Return)

xiv. Manson M, Logan W, Loy R. *Rubella and Other Virus Infections during Pregnancy*. London: Her Royal Majesty's Stationery Office; 1960. (↲ Return)

xv. Lundstrom R. Rubella during pregnancy: a follow-up study of children born after an epidemic of rubella in Sweden, 1951, with additional investigations on prophylaxis and treatment of maternal rubella. *Acta Paediatr Suppl*. 1962;133:1-110. (↲ Return)

xvi. Centers for Disease Control (CDC). Possible transfusion-associated acquired immune deficiency syndrome (AIDS)—California. MMWR *Morb Mortal Wkly Rep*. 1982;31(48):652-654. (↲ Return)

xvii. Centers for Disease Control (CDC). Pneumocystis pneumonia—Los Angeles. MMWR *Morb Mortal Wkly Rep*. 1981;30(21):250-252. (↲ Return)

xviii. Centers for Disease Control (CDC). Unexplained immunodeficiency and opportunistic infections in infants—New York, New Jersey, California. MMWR *Morb Mortal Wkly Rep*. 1982;31(49):665-667. (↲ Return)

xix. Centers for Disease Control and Prevention (CDC). Severe acute respiratory syndrome—Singapore, 2003. MMWR *Morb Mortal Wkly Rep*. 2003;52(18):405-411. (↲ Return)

xx. Centers for Disease Control and Prevention (CDC). Update: severe acute respiratory syndrome—United States, May 14, 2003. MMWR *Morb Mortal Wkly Rep*. 2003;52(19):436-438. (↲ Return)

xxi. Centers for Disease Control and Prevention (CDC). Cluster of severe acute respiratory syndrome cases among protected health-care workers—Toronto, Canada, April 2003. MMWR *Morb Mortal Wkly Rep*. 2003;52(19):433-436. (↲ Return)

xxii. Egger M, Smith GD, Altman DG, eds. *Systematic Reviews in Health Care: Meta-analysis in Context*. London: BMJ Publishing; 2001. (↲ Return 1) (↲ Return 2)

xxiii. Harris JD, Quatman CE, Manring MM, Siston RA, Flanigan DC. How to write a systematic review. *Am J Sports Med*. 2014;42(11):2761-2768. doi:10.1177/0363546513497567 (↲ Return)

xxiv. Gilbert R, Salanti G, Harden M, See S. Infant sleeping position and the sudden infant death syndrome: systematic review of observational studies and historical review of recommendations from 1940 to 2002. *Int J Epidemiol*. 2005;34(4):874-887. doi:10.1093/ije/dyi088

xxv. CDC. Safe sleep for babies. Centers for Disease Control and Prevention. 2018. https://www.cdc.gov/vitalsigns/safesleep/index.html. Accessed January 10, 2019

xxvi. Task Force on Sudden Infant Death Syndrome, Moon RY. SIDS and other sleep-related infant deaths: expansion of recommendations for a safe infant sleeping environment. *Pediatrics.* 2011;128(5):1030-1039.

xxvii. Carbine NE, Lostumbo L, Wallace J, Ko H. Risk-reducing mastectomy for the prevention of primary breast cancer. *Cochrane Database Syst Rev.* 2018;4:CD002748. doi:10.1002/14651858.CD002748.pub4 (↲ Return 1) (↲ Return 2)

10. Causality and Causal Thinking in Epidemiology

Learning Objectives

After reading this chapter, you will be able to do the following:

1. Discuss the 3 tenets of human disease causality
2. Explain how causal thinking plays a role in the epidemiology research process
3. Apply epidemiologic causal thinking to common exposure/disease problems

I have mentioned in previous chapters that it is difficult to use epidemiologic studies to "prove" that an exposure/disease association is causal. Randomized trials are occasionally an exception, and I discuss this further below. First, however, I will summarize various ways of thinking about causes of disease in humans, and then in the second half of the chapter, I will discuss how these causal theories apply to the epidemiologic literature specifically.

Causes of Human Disease

It is now a well-established idea that any given case of disease in a human is multifactorial. That is, there is no one specific cause per se but rather a multitude of factors that work in concert to cause a disease to begin. Various authors have described this basic concept with theoretical models: the sufficient component cause model (a.k.a. "causal pies"),[i] the social-ecologic model,[ii] and as the web of causation,[iii] among others. Though these models differ in their details, they share numerous common ideas in addition to multicausality, which is discussed below.

Tenets of human disease #1

All cases of disease have multiple causes.

First, a model for thinking about causes of disease, for those of you who think in pictures. Think of a jar, the kind in which you might serve drinks at a party:

Figure 10-1.
Source: *https://www.pinterest.com/pin/91057223695271823/*

Nonmodifiable characteristics of the person—genetics, family socioeconomic status while the person is young, and so on—determine the size of each jar (one for each disease) with which one starts. For example, someone with a high genetic risk of a certain disease starts with a smaller jar than someone without those genes. As the person moves through life, they encounter adverse exposures that add liquid to the jar; conversely, they can encounter protective exposures that drain liquid back out from the bottom spigot. In this model, the disease in question would begin when the jar is full to the top.

Using breast cancer as an example, the size of my "breast cancer jar" is determined by my genetics, the intrauterine environment in which I was a fetus (including anything my mother might have been exposed to while pregnant), my family's situation while I was growing up (including the laws and regulations that applied where we lived), and my (genetically determined) age at **menarche** and **menopause.** Then as I move through life, the fullness of my jar changes as I encounter detrimental and protective exposures. So every alcoholic drink adds a bit to the jar, as does any use of hormonal birth control, since these are associated with an increased risk of breast cancer. Conversely, every bout of physical activity and every pregnancy (both associated with reduced risks of breast cancer) take a bit out.[iv,v,vi]

Tenets of human disease #2:

Not all causes act at the same time

If I have a strong family history of breast cancer, and thus start life with a smaller jar, the number of adverse exposures I can withstand before cancer starts is much less than for someone without such a history (who has a bigger jar). Furthermore, each person might have a slightly different set of exposures that are either raising or lowering the level in their respective jars. While it is well-established at this point that smoking causes lung cancer,[vii] there are cases of lung cancer that arise in nonsmokers (so other exposures filled their jars), and there are lifelong smokers who nonetheless never develop lung cancer (their jars were likely big enough to start with that even thousands of cigarettes are not enough to fill it up).

Tenets of human disease #3:

There are many different ways a jar could get filled; there are many different collections of exposures that, taken together, can cause a case of disease.

The tenets of disease causality we have discussed thus far have a number of logical sequelae. First, as public health and clinical professionals, we do not need to identify all possible causes of a disease before taking action. If we are reasonably sure that smoking is a contributing cause of lung cancer in at least some people, then we will be able to prevent some cases of lung cancer if we can eliminate smoking as an exposure from at least some of the population. It does not matter that not all cases of lung cancer include smoking in their jars; it is enough to know that some of them do.

Disease Onset Timing

Once a person accumulates enough causes (their jar is full), their disease begins. We have no idea and no way of ever knowing, given current technologies, how many such causes are "enough"; possibly, it is different for each person. We consider a disease "caused" once it begins. Thus, when thinking about causes of disease in this way, a preventive factor is one that either prevents disease altogether or delays disease onset for some length of time. We often talk about "preventing death" from various causes; we cannot, of course, prevent death. We can only delay it.

Second, there are implications for the so-called strength of individual causes as well as **attributable fractions**. You will often hear people calling particular causes "strong" or "weak"—usually referring to the overall, population-level measure of association between the given exposure and the outcome. However, this idea only works if the prevalence of all causes in the population does not change. If we eliminate smoking, which in the US has odds ratios of around 40.0 when associating it with lung cancer,[viii] suddenly radon (ORs for radon-related lung cancer are usually more like 1.5 or 3.0[ix]) will look like a much stronger cause of lung cancer. Attributable fractions supposedly quantify the proportion of cases that were caused by—or can be "attributed to"—a particular exposure (see discussion of attributable risk under "Risk Difference" in chapter 4). However, because each case of disease has multiple causes filling its jar, the attributable fractions for all possible causes will sum to well over 100%, rendering this measure of association rather less than useful.

Determining When Associations Are Causal in Epidemiologic Studies

As mentioned in chapter 4, in epidemiology we look for evidence that exposures and outcomes are associated statistically. Epidemiologists are usually very careful not to use causal language. As you read studies from the epidemiology literature, you will see phrases like "associated with," "evidence in favor of," "possible," and other similar, carefully non-definitive semantics. However, sooner or later we must stop hedging and determine whether something is or isn't causal, because public health and clinical policy cannot be based on associations. What is the protocol for doing this? This schematic displays a possible flow chart for an epidemiologic research question:

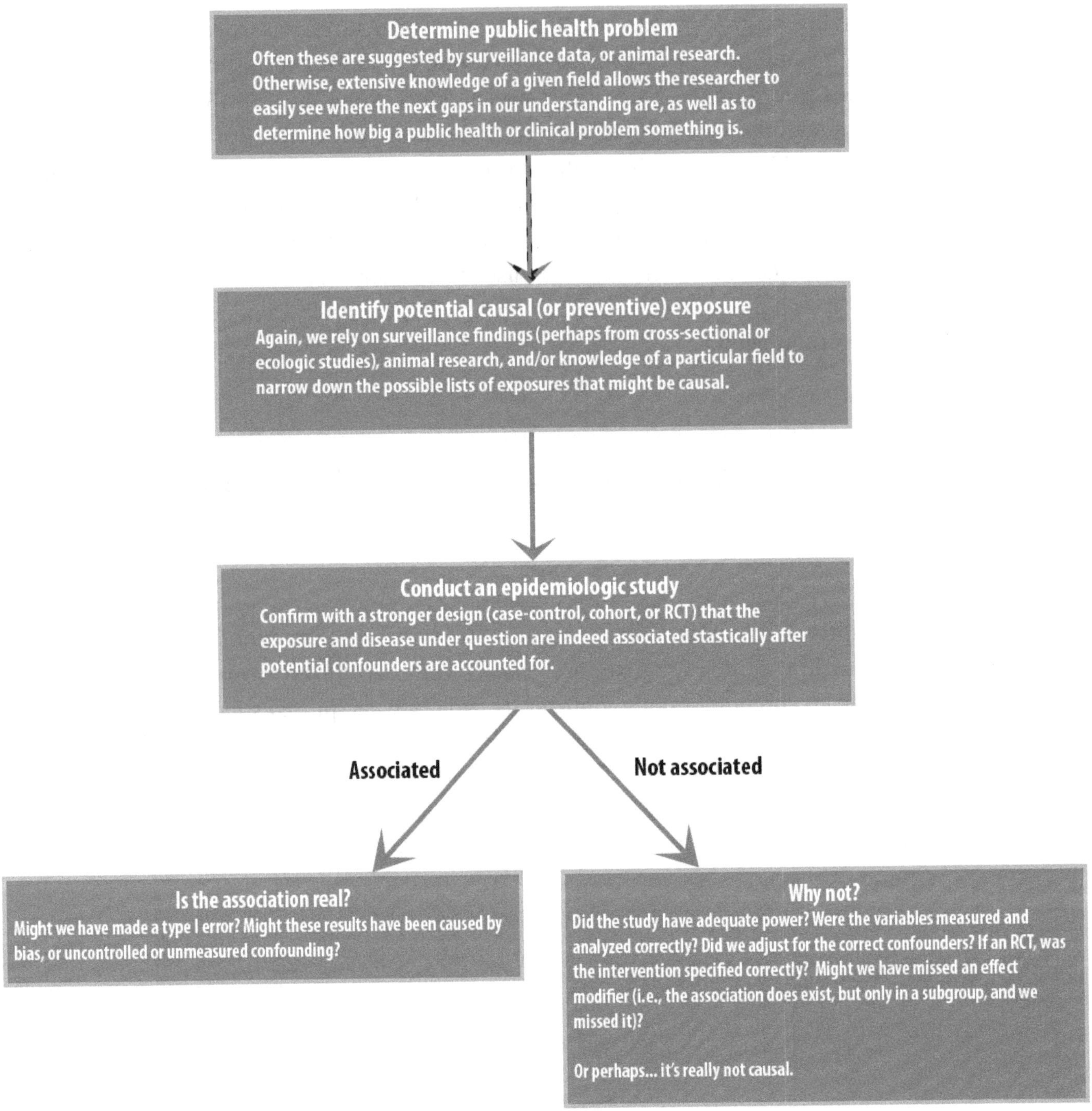

Figure 10-2

Only if we determine that the association is indeed real and not an artifact of bias, confounding, or random chance would we begin to assess whether or not it might be causal. This assessment requires a thorough understanding of the research question, of any underlying biology and physiology, and of previous work on the topic, and this assessment is not typically done by one person alone. Rather, we each do our studies, publish them, read other people's research, talk to each other at conferences, consult with colleagues from related disciplines, and so on. Slowly, collectively, the broad public health field moves toward a consensus for a given exposure/disease causal relationship.

There exist numerous checklist-style lists of criteria for determining whether an epidemiologic association is causal; the most famous are the "causal considerations" published by Sir Austin Bradford Hill—he was very careful not to call them criteria because they are just things to think about rather than a method for conclusively obtaining "the" answer.[x] For any given exposure/disease causal question, going through such lists is certainly a useful exercise, but I urge caution, as they are far from definitive in either direction. For instance, one of the items from Hill's article is "specificity," meaning that one cause leads to one effect. This works well for infectious diseases—HIV causes AIDS, but not also other things (although progression to full-blown AIDS requires causes in addition to HIV infection, such as lack of access to antiretroviral drugs)—but less well for chronic diseases. For instance, if we insist on meeting Hill's specificity criterion, smoking cannot be a cause of lung cancer because it also appears to cause heart disease, oropharyngeal cancer, and other outcomes. We know this to be untrue: smoking certainly causes lung cancer (and likely all the other conditions too). However, many of the other considerations on Hill's list do apply in the case of smoking: there is a dose-response association (more smoking correlates to a higher risk of lung cancer), there is biologic plausibility (cigarettes contain compounds known to be carcinogens), and there is consistency (all studies on the topic reach the same conclusion).

Methods & Considerations

Of all the (non-review) study designs we have considered, RCTs provide the best evidence in favor of causality, assuming these studies were correctly conducted and showed an association. This is because the randomization process, as discussed in previous chapters, renders confounding moot: if everything is the same between the 2 groups except for the intervention, then that intervention almost certainly is responsible for any difference in outcomes between the 2 groups. Indeed, one of Hill's considerations is whether there exists experimental (i.e., RCT) evidence on the topic. However, it is always possible that bias or random error is instead responsible for the results, as RCTs are not inherently free of these. Readers of randomized trials must use the same caution as readers of other types of studies and carefully evaluate the study's methods and results before drawing firm conclusions.

There are, however, numerous situations in which RCTs are not feasible or not ethical. For these research topics, in addition to the study design possibilities mentioned in chapter 7 (matching or enrolling a narrowly limited sample), there are a number of statistical methods that essentially aim to simulate a randomized trial using observational data. Examples of such methods include propensity score matching (which allows matching on dozens of variables at once, a feat that is not possible with conventional matching protocols) and inverse probability weighting (in which each "type" of observed participant—underweight, 80-year-old Black women with hypertension, perhaps—contributes to the final analysis according to how common that type of person is in the dataset and the target population). Doctoral students in epidemiology take entire courses on such causal inference methods; additional details are beyond the scope of this book, though interested students might consult a recent, introductory-level article series on this topic.[xi,xii,xiii]

Conclusion

Public health and clinical professionals rely on knowing whether a particular exposure causes a particular disease because intervention and policy changes depend on this knowledge. However, determining causality using epidemiologic research is a tricky proposition that relies on knowledge of underlying biology, physiology, and/or toxicology; awareness of any existing in vitro or animal studies; and careful readings of the existing epidemiologic literature on a given topic. All cases of disease have multiple causes, and these do not act simultaneously; each case of disease likely has a slightly different mix of contributing causes. However, we do not need to know all possible causes before taking action, as we can prevent some cases (stop some jars from filling) by intervening on even just a single known cause.

References

i. Rothman KJ, Greenland S. Causation and causal inference in epidemiology. *Am J Public Health.* 2005;95(suppl 1):S144-150. doi:10.2105/AJPH.2004.059204 (↲ Return)

ii. The social-ecological model: a framework for prevention. Center for Disease Control and Prevention (CDC). 2018. https://www.cdc.gov/violenceprevention/overview/social-ecologicalmodel.html. Accessed November 2, 2018. (↲ Return)

iii. Ventriglio A, Bellomo A, Bhugra D. Web of causation and its implications for epidemiological research. *Int J Soc Psychiatry.* 2016;62(1):3-4. doi:10.1177/0020764015587629 (↲ Return)

iv. Mørch LS, Skovlund CW, Hannaford PC, Iversen L, Fielding S, Lidegaard Ø. Contemporary hormonal contraception and the risk of breast cancer. *N Engl J Med.* 2017;377(23):2228-2239. doi:10.1056/NEJMoa1700732 (↲ Return)

v. McPherson K, Steel CM, Dixon JM. Breast cancer—epidemiology, risk factors, and genetics. BMJ. 2000;321(7261):624-628. (↲ Return)

vi. Tao Z, Shi A, Lu C, Song T, Zhang Z, Zhao J. Breast cancer: epidemiology and etiology. *Cell Biochem Biophys*. 2015;72(2):333-338. doi:10.1007/s12013-014-0459-6 (↲ Return)

vii.Health CO on S and. Smoking and tobacco use: history of the Surgeon General's Report. 2017. http://www.cdc.gov/tobacco/data_statistics/sgr/history/. Accessed October 30, 2018. (↲ Return)

viii. Stellman SD, Takezaki T, Wang L, et al. Smoking and lung cancer risk in American and Japanese men: an international case-control study. *Cancer Epidemiol Biomark Prev Publ Am Assoc Cancer Res Cosponsored Am Soc Prev Oncol*. 2001;10(11):1193-1199. (↲ Return)

ix. Zhang Z-L, Sun J, Dong J-Y, et al. Residential radon and lung cancer risk: an updated meta-analysis of case-control studies. *Asian Pac J Cancer Prev* APJCP. 2012;13(6):2459-2465. (↲ Return)

x.Hill AB. The environment and disease: association or causation? *Proc R Soc Med*. 1965;58:295-300. (↲ Return)

xi. Snowden JM, Tilden EL. Further applications of advanced methods to infer causes in the study of physiologic childbirth. *J Midwifery Womens Health*. 2018;63(6):710-720. doi:10.1111/jmwh.12732 (↲ Return)

xii. Snowden JM, Tilden EL, Odden MC. Formulating and answering high-impact causal questions in physiologic childbirth science: concepts and assumptions. *J Midwifery Womens Health*. 2018;63(6):721-730. doi:10.1111/jmwh.12868 (↲ Return)

xiii. Tilden EL, Snowden JM. The causal inference framework: a primer on concepts and methods for improving the study of well-woman childbearing processes. *J Midwifery Womens Health*. 2018;63(6):700-709. doi:10.1111/jmwh.12710 (↲ Return)

11. Screening and Diagnostic Testing

Learning Objectives

After reading this chapter, you will be able to do the following:

1. Differentiate between screening and diagnostic testing
2. Calculate and interpret common test characteristics
3. Discuss the role of public health in screening programs

Abbreviations used in this chapter

Dx = disease
Sx = symptoms
Hx = history
Tx = treatment
Sn = sensitivity
Sp = specificity
PPV = positive predictive value
NPV = negative predictive value
Pt = patient.

Introduction

In this chapter, we will cover both **screening** and **diagnostic testing**. Though public health professionals are not usually directly involved with diagnosing patients, the tests used for screening and diagnostic testing are often the same (the difference being context), and the same mathematical tools are used to assess the accuracy of these tests. In addition, public health professionals who are involved in disease surveillance may need to know how to interpret these results when evaluating surveillance case definitions.

Screening versus Diagnostic Testing

The word *screening* refers to testing an *asymptomatic* population for a particular condition in order to identify those who have the condition so that they can be treated early. Common screening tests currently used in the US include various cancer screenings (mammograms, pap smears, skin checks for those at high risk of melanoma); routine **hypertension** screening at doctors' offices (this is why they take your blood pressure every time you go); hearing, vision, and dental screening at elementary schools; annual tuberculosis and HIV screening among health care workers, and so on. Public health officials are often involved in screening programs either directly (providing personnel to go to the elementary school to conduct the hearing screenings) or indirectly (health education campaigns to increase the use of pap smears).

Diagnostic testing, on the other hand, is performed on a patient who is symptomatic in order to determine what condition they have. Clinicians perform what is called differential diagnosis when confronted with a patient with new complaints. In a nutshell, the doctor, nurse practitioner, or other health care provider takes all known information from the patient's history and physical exam and decides what could be wrong. For instance, if a 24-year-old female presents to the clinic complaining of visual disturbances followed by severe headache, this could be any number of things: concussion, migraine with aura, hemorrhagic stroke[1], meningitis, and so on. The clinician's task is to determine which one it is so that the patient can be treated correctly. The differential diagnosis process involves administering diagnostic tests that are designed to either rule in or rule out conditions on the differential diagnosis list.

Returning to our example, the likelihood of concussion could be assessed by questioning the patient about any recent head or neck trauma (questions can be diagnostic tests!). If the patient denies such trauma (e.g., she has not played a contact sport, fallen, been in a motor vehicle accident, etc. in the last 24 hours), then we have probably ruled out concussion. The next task would be to rule out hemorrhagic stroke— strokes in young people are very rare but not unheard of, and the faster they are treated, the better the **prognosis**. We could rule out hemorrhagic stroke by looking for blood in the patient's cerebrospinal fluid (the diagnostic test in that case is a spinal tap). Assuming that the patient's spinal fluid is indeed clear of blood, we would then test to rule

1. There are two kinds of strokes: ischemic and hemorrhagic. Ischemic strokes are caused by blood clots in the brain, and hemorrhagic strokes are caused by bleeding in the brain. The latter are much more common in young people (ischemic strokes would be almost unheard of in an otherwise-healthy 24-year-old female), though ischemic strokes are more prevalent overall. These are really two different diseases, but they produce a very similar set of symptoms, so prior to our understanding the disease processes behind them, all such symptoms were classified as "stroke."

out meningitis. If that test is also clear, then we could safely assume that she has a migraine and offer treatment accordingly.

The order in which one rules in or rules out conditions on the differential diagnosis list depends on their relative severity, the costs associated with various tests, and the prevalence of the conditions in question. In this scenario, it was very important to quickly rule out stroke, because if it were a stroke, treatment would need to be initiated as soon as possible—whereas a migraine can wait an hour without causing further harm. This isn't pleasant for the person with the migraine, but it won't kill them or cause long-term disability. A brain tumor is also on the differential diagnosis list; however, that condition is very rare in patients in their 20s, and delaying treatment for a brain tumor by 24 hours would not matter (unlike for a stroke, which is also rare but for which one cannot delay treatment). Thus after ruling out concussion, meningitis, and stroke, we would treat the patient for a migraine. If she did not get better in 24–48 hours, we would return to the differential diagnosis list and possibly test for neuroblastoma or other brain cancers.

As mentioned above, often the same tests are used both for screening and for diagnostic testing; the distinction depends on context. If I do not have any symptoms of breast cancer and have a mammogram, then it is a *screening test* (screening is done in asymptomatic populations). If on the other hand I find a lump in my breast, go to the doctor, and she sends me for a mammogram, it is now a *diagnostic test*, because I am symptomatic.

Disease Critical Points and Other Things to Understand about Screening

The figures in this section and the idea of critical points are adapted from lectures given by Dr. David Slawson at the University of Virginia Medical System, Department of Family Medicine, in 2004-2005.

The natural course of a medical condition looks like this:

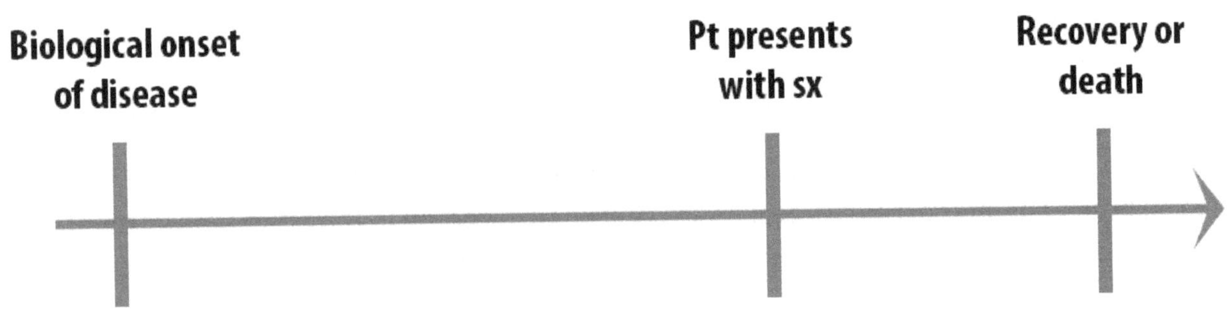

Figure 11-1

The first step is the biological onset of the disease. This could be the first mutation that turns the cell into a cancerous cell. It could be a virus getting in through someone's mucous membranes and beginning to replicate. Importantly, biological onset is not observable.

Eventually, the person with the disease will have symptoms severe enough that they seek treatment: they go to a clinic, they go to the emergency room, they go to the pharmacy and buy some decongestant; they will eventually seek treatment of some kind. The exact timing will vary from person to person and from disease to disease.

The final stage in our natural history of disease is the outcome. Either they get better or they don't.

So then what is screening? This is the idea of screening:

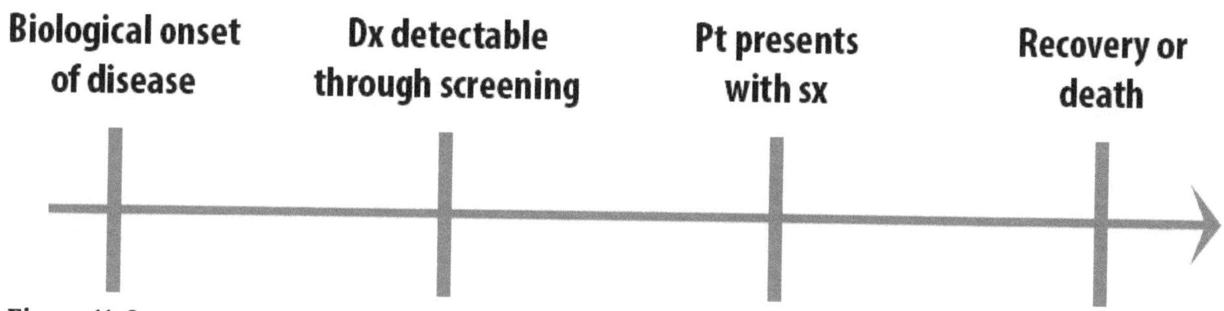

Figure 11-2

Remember that a screening program starts with asymptomatic people — normal, everyday people — and tests them to see if they have the disease.

The idea is to find the disease early. *Screening* __*is not*__ *primary prevention.*[2] Screening finds early disease; it does not prevent the disease from occurring. Screening might well prevent poor outcomes from the disease (secondary prevention), but it doesn't prevent the disease itself.

Thus the first criterion for a successful screening program is that a test exists that can detect early, pre-symptom disease. This is not always the case — for instance, we don't screen for ovarian cancer partly because no such test exists at this time. A second criterion for a successful screening program is that the condition is prevalent enough and/or the costs of not treating it are high enough to make it worth screening for on a population level. Another reason we don't screen for ovarian cancer is because the prevalence is very low, and thus a screening program would arguably not be worth the resources.

2. This mistake regularly appears in the literature! See for instance Ring et al.[i]

The next important idea is that every disease has a critical point. If you treat the disease before it gets to the critical point, you can make a difference in the outcome: the patient will be cured, will live longer, or will live better. However, if you treat the disease after the critical point, the treatment will have no effect (this is why we have hospice—after a certain point, further aggressive treatments are not useful and are potentially even harmful).

When thinking of implementing a screening program for a given disease, one must consider the timing of the critical points. For instance, what if the critical point (represented by the red line) for a given disease occurs here:

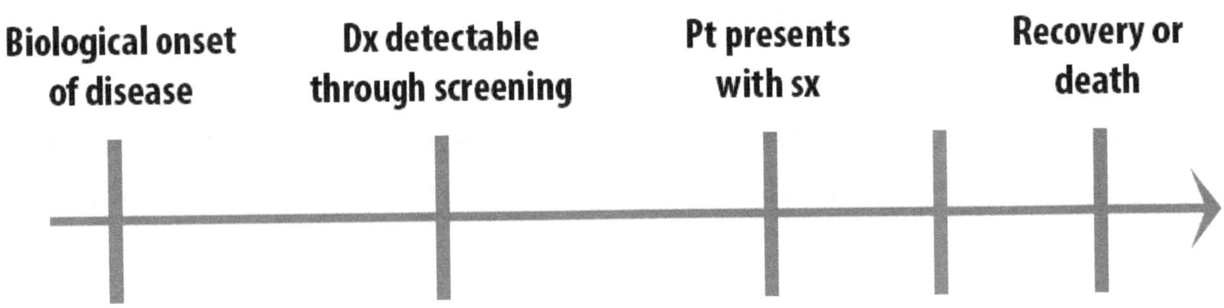

Figure 11-3

In the above scenario (Figure 11-3), it would not be cost-beneficial to screen for this condition, because detecting it early doesn't help. By the time people seek help for their symptoms, there is still plenty of time to treat them. Screening in this scenario would not cause physical or emotional harm per se, but it would waste resources. We are learning that both prostate cancer and breast cancer probably fall into this category. Except for certain very high-risk groups, most men are no longer routinely screened for prostate cancer, and there is increasing evidence that mammograms might only be truly useful for high-risk women.[ii]

What if instead the critical point is here:

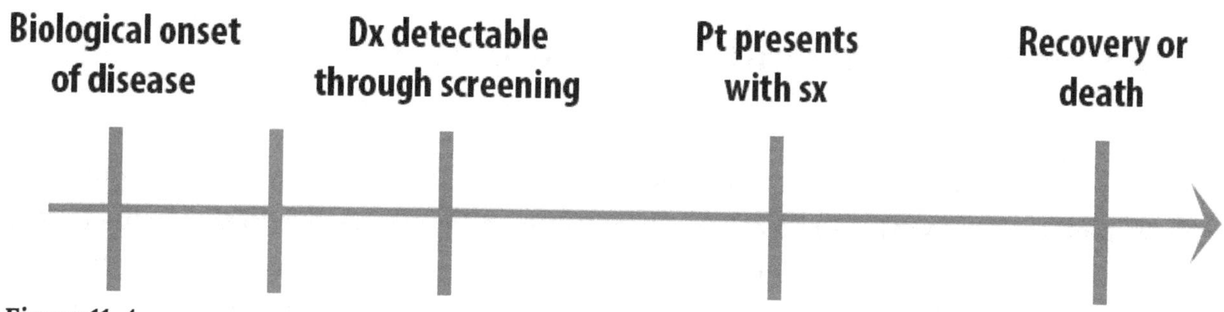

Figure 11-4

In this case (Figure 11-4), we also probably would not screen for this condition because by the time we screen, it's already too late. Screening in this scenario would cause emotional harm because people would know that they have an untreatable disease for a longer period of time. For a highly contagious condition, however, we might screen people in this scenario—not so that we can treat them but so that they can take precautions and not spread it to others. Before we had antiretroviral drugs, for instance, we screened high-risk populations for HIV.

Screening, then, is most useful in this scenario:

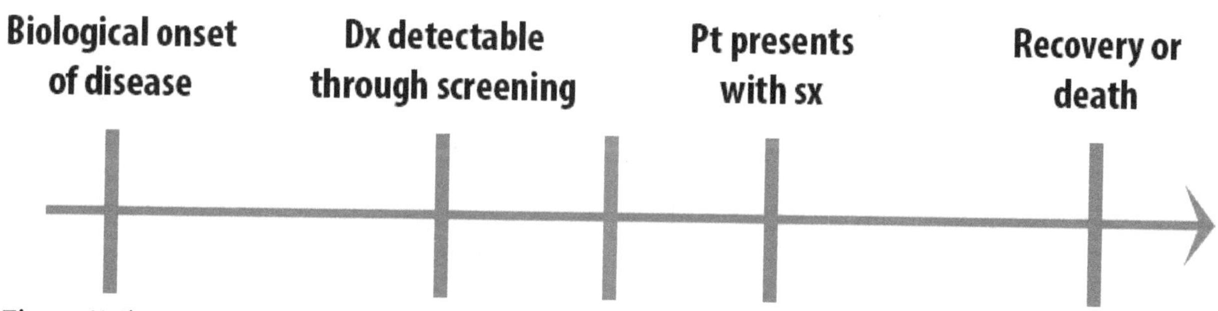

Figure 11-5

In this case, early detection *does* make a difference. By the time some patients present with symptoms, it is too late to treat them, but if we can detect the disease earlier than that, there is still time to help them. We still might not implement a population-wide screening program, if the disease is rare and the cost of the screening test quite high, for instance. But only in this scenario, where the critical point lies between screening detection and treatment seeking, should we even consider screening.

Accuracy of Screening and Diagnostic Tests

There are 4 **test characteristics** that we use to quantify how accurate a particular test is: **sensitivity**, **specificity**, **positive predictive value (PPV)**, and **negative predictive value (NPV)**. The first 2 are known as "fixed test characteristics" because they do not change, regardless of disease prevalence. The PPV and NPV, however, do change when disease prevalence in the underlying population changes.

Calculation of these 4 test characteristics requires that we arrange our data in a 2 × 2 table. This time, however, instead of exposure on the left, we have the test result on the left:

Table 11.1

	D+	D-	Total
T+	**TP**	**FP**	TP+FP
T-	**FN**	**TN**	FN+TN
Total	TP+FN	FP+TN	**TP+FP+TN+FN**

Looking at the screening 2 × 2 table, you'll notice that the cells are no longer labeled ABCD. They still could be—and indeed, in many textbooks they are. However, I prefer this notation because it reminds you what each cell is in this scenario. Specifically, the top left cell contains "true positives"—these individuals do have the disease (according to a gold standard diagnosis method–see below) and tested positive using the screening or diagnostic test. The top right cell, on the other hand, contains "false positives"—people who tested positive using the screening or diagnostic test but do not actually have the disease. The bottom left cell contains the "false negatives"—these individuals have the disease but for some reason test negative. Finally, the bottom right cell is comprised of the "true negatives," who do not have the disease and test negative.

One obtains data for a screening 2 × 2 table by administering both the test that we're evaluating and a **gold-standard** diagnostic method to a large group of people. For example, the gold standard for diagnosing Alzheimer's disease is the presence of a certain type of brain plaque, observable upon autopsy. Since this is not a feasible thing to do to living elderly patients who have suspected dementia, researchers developed the Mini Mental State (MMS) test.[iii] To create the above 2 × 2 table, then, one would administer the MMS to a group of elderly persons and categorize their results as either testing positive (T+) or testing negative (T−). Then one would collect autopsy data as they die and categorize those same individuals as either having had Alzheimer's (D+) or not (D−).

In other scenarios, both can be done at the same time (rather than waiting for death)—for example, using the Beck Depression Inventory (BDI) as a test for depression and comparing results against a series of visits with a mental health professional qualified to definitively diagnose depression. The reason for having a test in the latter case is that the BDI is much quicker and cheaper than sending everyone for a formal psychiatric evaluation; the BDI is also a self-administered questionnaire, which can be utilized in large cohort studies (whereas the clinician-mediated diagnosis is untenable for large and/or geographically disparate study populations).

Note that one can calculate prevalence of the disease in the sample using a screening 2×2 table: everyone with disease (TP+FN) over everyone in the sample.

Sensitivity and Specificity

Sensitivity (Sn) is the probability that a patient tests positive given that they have the disease. In probability notation, this is written as follows:

$$Sn = P(T+|D+) = \frac{TP}{(TP+FN)}$$

Specificity (Sp) is the probability that a patient tests negative given that they do not have the disease:

$$Sp = P(T-|D-) = \frac{TN}{(FP+TN)}$$

Both of these values are proportions and are usually expressed as percentages.

Recall that sensitivity and specificity are fixed test characteristics—they do not change if the prevalence of the condition in the sample changes. Sensitivity and specificity values are published when new tests become available, and are used by clinicians to decide what tests to order.

For diagnostic testing, recall that we are trying sometimes to *rule in* and other times to *rule out*. Two common mnemonic devices are *SpIN* and *SnOUT*: a test with high specificity, when positive, rules IN, and a test with high sensitivity, when negative, rules OUT. To understand why, look back at the formulae. The denominator for specificity is FP+TN (all the individuals without the disease), and the numerator is just TN. If the specificity is high (close to 100%), then FP must be very low. Thus a patient with a positive result on a highly specific test is probably a true positive. The same logic can be used to understand SnOUT: the denominator for sensitivity is TP+FN (all the individuals with the disease), and the numerator is just TP. If sensitivity is near 100%, then by definition there are few false negatives. A negative result on a highly sensitive test is thus almost certainly a true negative. Thus clinicians will choose a test with high sensitivity when they want to rule out (as in concussion or stroke, in our example above), and a test with high specificity when they want to rule in.

For screening purposes, we are testing an asymptomatic population—we thus want to minimize false negatives. This is because we wouldn't want to tell someone that they are disease-free if they're really not. Screening programs therefore utilize tests with high sensitivities.

Positive and Negative Predictive Values

The positive predictive value is the probability that you actually have the disease given that you tested positive (look carefully, and be sure you understand how this is different than sensitivity):

$$\text{PPV} = P(D+|T+) = \frac{TP}{(TP+FP)}$$

The negative predictive value is the probability that you do not have the disease given that you tested negative:

$$\text{NPV} = P(D-|T-) = \frac{TN}{(FN + TN)}$$

Again, these quantities are proportions and are usually expressed as percentages.

PPV and NPV are used to interpret test results once those results are known. Unlike sensitivity and specificity, however, PPV and NPV do change as the prevalence of disease in the sample changes—it is thus important to know something about the prevalence of disease in the target population to which an individual belongs before you can interpret their test results. For example, if a patient tests positive for tuberculosis (TB) and you know that the prevalence of TB in the population from which the patient comes is 10%, and the PPV given a 10% prevalence is 52.6%, then the interpretation of that test result is: there is a 52.6% chance that this patient has TB (there is thus a 47.4% chance that they do not have TB and the result was a false positive).

In general, PPV will decrease and NPV will increase as prevalence decreases. This makes intuitive sense: as a condition becomes more rare, then guessing that the patient does not have the disease becomes more and more likely to be correct. Extremely low PPV secondary to low prevalence was the rationale behind the 2009 recommendation by the US Preventive Services Task Force (USPSTF) that women in their 40s stop being screened for breast cancer unless they are extremely high-risk.[iv] The prevalence of disease among women in their 40s is very low (0.98%)[v] and the PPV in this population is thus well under 1%—this means that greater than 99% of women who are sent for follow-up testing (breast biopsy, usually) are false positives and thus undergo this expensive, invasive follow-up (with its corresponding emotional stress) unnecessarily. For women with a strong family history and/or who are known to be BRCA-1 or BRCA-2 carriers, mammography for 40-year-olds is still warranted—because these women come from an underlying population in which the prevalence (and thus the PPV) is much higher.

Example

Say a new test for anemia is developed that does not require a finger stick to obtain blood (no one likes needles!)—perhaps using a scanner that can detect hemoglobin levels through the thin skin on the underside of a wrist. The following 2 × 2 table is published:

Table 11.2

	D+	D-	Total
T+	101	15	116
T-	18	866	884
Total	119	881	**1000**

In this case, the test results (either T+ or T-) would come from the wrist scanner, and the disease results (D+ or D-) would come from the usual method of diagnosing anemia, which requires a blood draw.

Sample Calculations

Using the data from the above table, we can calculate the four test characteristics.

$$\text{Sensitivity} = \frac{101}{(101 + 18)} = 84.9\%$$

$$\text{Specificity} = \frac{866}{(866 + 15)} = 98.3\%$$

$$\text{Positive Predictive Value} = \frac{101}{(101 + 15)} = 87.0\%$$

$$\text{Negative Predictive Value} = \frac{866}{(866 + 18)} = 98.0\%$$

We can also calculate the prevalence of anemia in this sample.

$$\text{Prevalence} = \frac{(101 + 18)}{(101 + 18 + 15 + 866)} = 11.9\%$$

Let's now say that there is a patient who took the new test and tested positive. However, we know that the patient is from a population with a lower prevalence of anemia than 11.9%—adolescent males, for example, in whom the prevalence is around 1%.[vi] In this case, the above PPV no longer applies. However, since we know the Sensitivity and Specificity, we can create a new 2 × 2 table, from which we can calculate a new PPV for a lower prevalence population. We begin by deciding (arbitrarily) that we will again have 1,000 people in the table:

Table 11.3

	D+	D-	total
T+			
T-			
total			**1000**

If the prevalence is 1%, then 10 people would be expected to have the disease:

Table 11.4

	D+	D-	total
T+			
T-			
total	10	990	**1000**

Since the sensitivity is 84.9%, and sensitivity is a fixed test characteristic, we can solve for true positives as follows:

$$Sn = \frac{TP}{(TP + FN)}$$

0.849 = TP/10

TP = 8.49

We then subtract to get false negatives:

Table 11.5

	D+	D-	Total
T+	8.49		
T-	1.51		
Total	10	990	1,000

We can do a similar calculation with the known specificity of 98.3%:

(0.983)(990) = TN = 973.17

We can then fill in the nondiseased column:

Table 11.6

	D+	D-	Total
T+	8.49	16.83	25.32
T-	1.51	973.17	974.68
Total	10	990	1,000

Now we can calculate our new PPV:

PPV = 8.49/25.32 = 33.5%

There is thus only a 33.5% chance that our male adolescent actually has anemia based on a positive skin scan test. (We would follow up with a blood draw to confirm—had he tested negative, the new NPV would have been 99.9%—no blood draw required!)

Summary

Screening and diagnostic testing are similar procedures; the difference depends on context (whether the tested person is symptomatic or not). Accuracy of such tests is quantified by sensitivity and specificity (used ahead of time to pick the correct test), and positive and negative predictive values (used after the test results are known, to interpret them). One must know the prevalence of a disease in the target population in order to use PPV and NPV.

References

i. Ring KL, Modesitt SC. Hereditary cancers in gynecology: what physicians should know about genetic testing, screening, and risk reduction. Obstet Gynecol Clin North Am. 2018;45(1):155-173. doi:10.1016/j.ogc.2017.10.011 (↲ Return)

ii. Welch HG, Prorok PC, Kramer BS. Breast-cancer tumor size and screening effectiveness. N Engl J Med. 2017;376(1):94-95. doi:10.1056/NEJMc1614282 (↲ Return)

iii. Rovner BW, Folstein MF. Mini-mental state exam in clinical practice. Hosp Pract Off Ed. 1987;22(1A):99, 103, 106, 110. (↲ Return)

iv. Final update summary: breast cancer, screening. US Preventive Services Task Force. https://www.uspreventiveservicestaskforce.org/Page/Document/UpdateSummaryFinal/breast-cancer-screening. Accessed November 27, 2018. (↲ Return)

v. USCS Data Visualizations. CDC. https://www.cdc.gov/cancer/dcpc/data/. Accessed November 27, 2018. (↲ Return)

vi. Looker AC, Dallman PR, Carroll MD, Gunter EW, Johnson CL. Prevalence of iron deficiency in the United States. JAMA. 1997;277(12):973-976. (↲ Return)

Appendix 1: How to Read an Epidemiologic Study

Key Takeaways

A standard epidemiology study (not counting the abstract—more on this later) has 4 parts:

- Introduction
- Methods
- Results
- Discussion

Usually these are labelled, but not always. Sometimes they have different labels (eg, "background" instead of "introduction.") Even without labels, epidemiology papers are almost always organized in this order. Details about each section are discussed below.

Introduction

The INTRODUCTION usually consists of three things:

- What we already know about a topic
  - A very select summary of what we know! It is important to remember that intros are NOT exhaustive literature reviews. Furthermore, what things are included is entirely at the authors' discretion (with some input from peer reviewers and editors), which means that you do see the occasional biased/incomplete introduction.
- What we don't know about the topic (ie, what is the gap in the literature?)
- What this study will do to address that gap
  - Usually concluding with "our study question was…" or "our objective here was…"

The introduction is where you will find answers to questions like "What is the public health or clinical problem this study is trying to address?" and "What was their research question?"

Methods

The METHODS is just that—a description of the methods used for this study. Ideally, the methods section will describe:

- How they got their sample from the target population/what dataset was used
  - Including inclusion/exclusion criteria, with rationales as appropriate
- What is the study design (including design-specific relevant details, such as how participants were randomized, if it's a randomized controlled trial)
  - Occasionally, if a study has been done using a well-known dataset (e.g., the NHANES data—see Chapter 3), the methods section will just direct the reader to other publications in which these methods are described in detail, rather than re-printing all of the information
- What was the exposure, how and when was it measured, and how was it operationalized in the analysis
  - ie, did they ask people their ages, but then dichotomize into "old" (>65) vs. "young" (65 and younger)?
- What was the outcome, how and when was it measured, and how was it operationalized in the analysis
  - ie, what was the case definition used for diagnosis? Were cases identified via clinics, or self-report, or some other method?
- What confounders and/or effect modifiers were included, how they were chosen, how they were measured, and how they were operationalized in the analysis
  - Collectively these are referred to as "covariables"
  - Any variables listed under "adjusted for" or "included in the model" are confounders
  - Any variables listed as "interactions" or "stratified by" are effect modifiers
- The statistical methods used

The methods section also should include a sentence about ethics/IRB approval, and informed consent, if applicable.

As a beginning epidemiology student, do not be concerned if you do not understand everything in the methods section! This is particularly true if the study included laboratory assays (e.g. to measure blood lead levels), but also pertains to the statistical methods. Papers must include enough detail in the methods section so that other scientists can evaluate, and potentially replicate, the work—which means they are written for other epidemiologists who are publishing papers, all of whom potentially have many years' worth of training in relevant methods.

Your task as a first-time epidemiology student (or as an end-user of epidemiologic research who has some – but not a lot of – training in the field) is to read the methods carefully enough to spot any potential sources of bias, given your level of understanding. For instance, after reading this book, my hope is that you could spot egregious selection bias by reading the authors' methods and thinking through "who did they get, who did they miss?". However, I would not expect that you would be able to spot a bias introduced because the authors violated one of the assumptions of the statistical model that they used. Sometimes I read papers myself where I don't quite follow the methods, particularly for laboratory-based measurements. In those cases, I just trust the peer review process—several pairs of eyes were on any given study before mine, so probably the methods are kosher. If it seems like maybe there's a problem, I ask one of my laboratory (or statistics, or clinical, depending on what my question is) colleagues about it.

Bottom line: read the methods carefully, but if there are parts you don't quite follow, don't worry about it unless you are going to cite that work yourself. In that case, ask around and find someone who can help you interpret the methods.

Results

The RESULTS section contains...results. What did they find? This section is usually very dense in terms of numbers. There will be odds ratios, risk ratios, confidence intervals, p-values, etc. Usually the results section begins with a discussion of the sample that was in the study, and this often further includes a table of demographics and relevant risk factors (usually "Table 1"). This table is a good place to get a feel for who was in the study (and therefore who was not). Then usually the authors will discuss the MAIN results: what did they find pertaining to their primary research question? They will not say what the results mean in this section (that's for the "discussion," below)—this section just presents the numbers. Expect to go back and forth between the text and the tables and the figures several times—most journals expressly ask authors not to duplicate results (meaning, if the results are presented in a table, don't repeat them in the text). Thus, to understand all of the results yourself, you will need to read both the text and the tables/figures. The last few paragraphs of the results section are used to present subgroup analyses, or bias/sensitivity analyses.

As you read results sections, think about what you read in the methods section. Do you believe these results, given the methods used?

Discussion

The DISCUSSION section is the *authors' opinions* about what the results mean. It usually begins with a summary of the main findings, and then compares these findings to other published findings on the same or similar topics. It should include a limitations (or strengths and limitations) sub-section, in which the authors provide a very frank picture of what limitations their study had (where there might have been bias, etc). If, when reading the methods and results sections, you thought of a potential bias that is NOT discussed here...pause. Perhaps this study is not the best source, then? All limitations should be acknowledged. (Corollary: no study is perfect! All have limitations. We can try to minimize, but need to 'fess up to the ones that remain.) The discussion section usually concludes with some kind of recommendation, either for policy, or further research. Again, this is the authors' opinion. Indeed, you are welcome to disagree entirely with any or all of a given discussion section—discussion sections are opinion, not fact.

Abstract

Finally, there is the ABSTRACT. Usually found at the beginning of the paper, often in its own separate box—this is a brief summary of the entire thing. Sometimes these same subheadings will be in the abstract, other times not. **WARNING: You cannot understand a paper just by reading the abstract.** Often only one or two main results are presented in the abstract, and the methods are quite sparse, as abstracts are limited usually to a few hundred words. *Never cite a paper if all you have read is the abstract.* This will come back to haunt you.

A few other details

Below are a few more points for if/when you are looking for papers yourself. Consider these things as you determine whether it's worth reading and/or citing a paper that you have found.

- Just under the paper's title is a list of the authors, their affiliations, and (usually) their credentials. Are these the kinds of people who need to be on this study? For instance, if a study on appropriate treatment for congestive heart failure does not include a cardiologist as an author, maybe that's a problem. If a study includes fancy statistical methods beyond basic logistic or linear regression, but the author list includes only clinicians, and no one with specialized statistical training, maybe that's a problem.
- Somewhere, usually on either the first or last page, or sometimes between the conclusion and the reference list, is a note about funding and other conflicts of interest. These are often

illuminating. For instance, I know of a study casting doubt on the benefits of breastfeeding that was funded by the International Formula Council.[i]

- Many journals will list dates—the date the article was submitted, the date it was received in revised form (meaning, it was received, sent out for review, the reviews were sent back to the authors, who then made the changes), and the date it was accepted. If the initial submission date and the acceptance date are not at least six weeks (and more realistically six months) apart, then it's possible that the peer review process was circumvented in some way. This happens. Sad but true.

References

i. Cope MB, Allison DB. Critical review of the World Health Organization's (WHO) 2007 report on "evidence of the long-term effects of breastfeeding: systematic reviews and meta-analysis" with respect to obesity. *Obes Rev Off J Int Assoc Study Obes.* 2008;9(6):594-605. doi:10.1111/j.1467-789X.2008.00504.x (↲ Return)

Appendix 2: Glossary

2x2 Table

A convenient way for epidemiologists to organize data, from which one calculates either *measures of association* or *test characteristics*.

Absolute measure of association

A *measure of association* calculated fundamentally by subtraction. See also *risk difference*.

Absolute risk

See *Incidence*.

Attributable fraction

A misleading *measure of association* that supposedly quantifies the proportion of cases of disease that can be "attributed" to a particular exposure. However, since every case of disease has more than one cause, the attributable fractions for all relevant exposures will sum to well over 100%, making the attributable fraction uninterpretable.

Baseline

The start of a *cohort study* or *randomized controlled trial*.

Bias

Systematic error. *Selection bias* stems from poor sampling (your sample is not representative of the target population), poor response rate from those invited to be in a study, treating cases and controls or exposed/unexposed differently, and/or unequal loss to follow up between groups. To assess selection bias, ask yourself "who did they get, and who did they miss?"--and then also ask yourself "does it matter"? Sometimes it does, other times, maybe it doesn't.

Misclassification bias means that something (either the exposure, the outcome, a confounder, or all three) were measured improperly. Examples include people not being able to tell you something, people not being willing to tell you something, and an objective measure that is somehow systematically wrong (eg always off in the same direction, like a blood pressure cuff that is not zeroed correctly). *Recall bias*, social desirability bias, interviewer bias--these are all examples of misclassification bias. The end result of all of them is that people are put into the wrong box in a *2x2 table*. If the misclassification is equally distributed between the groups (eg, both exposed and unexposed have equal chance of being put in the wrong box), it's *non-differential misclassification*. Otherwise, it's *differential misclassification*.

Case-control study

An *observational study* that begins by selecting cases (people with the disease) from the *target population*. One then selects controls (people without the disease)–importantly, the controls must come from the same target population as cases (so, if they suddenly developed the disease, they'd be a case). Also, selection of both cases and controls is done without regard to exposure status. After selecting both cases and controls, one then determines their previous exposure(s). This is a retrospective study design, and as such, more prone to things like *recall bias* than prospective designs. Case-control studies are necessary if the disease is rare and/or if the disease has a long *induction period*. The only appropriate measure of association is the *odds ratio*, because one cannot measure *incidence* in a case-control study.

Censored time

Time during which a given person is not contributing *person-time at risk* to a *cohort study* or *randomized controlled trial*. Left censoring happens before the person begins to contribute person-time at risk (because they are not yet enrolled in the study, even though the study has started), and right censoring happens after a person stops contributing person-time at risk (because they experienced the event of interest, a *competing risk*, or were lost to follow-up).

Cohort study

An observational design. Usually prospective, in which case one selects a *sample* of at-risk (non-diseased) people from the *target population*, assesses their exposure status, and then follows them over time looking for *incident cases* of disease. Because we measure *incidence*, the usual measure of association is either the *risk ratio* or the *rate ratio*, though occasionally one will see *odds ratios* reported instead. If the exposure under study is common (>10%), one can just select a sample from the target population; however, if the exposure is rare, then exposed persons are sampled deliberately. (Cohort studies are the only design available for rare exposures.) This whole thing can be done in a retrospective manner if one has access to existing records (employment or medical records, usually) from which one can go back and "create" the cohort of at-risk folks, measure their exposure status at that time, and then "follow" them and note who became diseased.

Comorbidity/comorbid condition

If a person has more than one disease at a time, all such diseases for that person are known as comorbidities or comorbid conditions.

Competing risks

In a *cohort study* or *randomized controlled trial*, competing risks are defined as "everything else that might kill someone or otherwise make them no longer at risk of the outcome under study." So, if we are studying ovarian cancer, then possible competing risks are fatal motor vehicle accidents, fatal heart attacks, etc., as well as oophorectomy (surgical removal of the ovaries). If someone experiences a competing risk, they no longer contribute *person-time at risk*.

Confidence interval

A way of quantifying *random error*. The correct interpretation of a confidence interval is: if you repeated the study 100 times (go back to your *target population*, get a new *sample*, measure everything, do the analysis), then 95 times out of 100 the confidence interval you calculate as part of this process will include the true value, assuming the study contains no *bias*. Here, the true value is the one that you would get if you were able to enroll everyone from the population into your study--this is almost never actually observable, since populations are usually too large to have everyone included in a sample. Corollary: If your population is small enough that you can have everyone in your study, then calculating a confidence interval is moot.

Confounding

A systematic error in a study (some people call it a *bias*; I prefer not to) that is caused by a third variable interfering in the exposure-disease relationship.

Count

A measure of disease frequency used in lieu of *prevalence* when the disease is extremely rare.

Cross-sectional study

An *observational study design* in which one takes a *sample* from the *target population*, assesses their exposure and disease status all at that one time. One is capturing *prevalent cases* of disease; thus the *odds ratio* is the correct measure of association. Cross-sectional studies are good because they are quick and cheap; however, one is faced with the chicken-egg problem of not knowing whether the exposure came before the disease.

Cumulative incidence

See *Incidence Proportion*.

Descriptive epidemiology

A summary of what is known about a particular condition, including data on *incidence*, *prevalence*, and known *risk factors*.

Determinants

Things that cause or prevent disease. Also called "causes."

Diagnostic testing

Applying a clinical test to a person who has presented with symptoms, to aid in determining what condition the person has, so that they can be correctly treated.

Differential misclassification bias

Misclassification that occurs in one study group more than another. Adversely affects *internal validity*.

Disproportionately distributed

Refers to a situation wherein exposed individuals have either more or less of the disease of interest (or diseased individuals have either more or less of the exposure of interest) than unexposed individuals.

Ecologic fallacy

A logical error that stems from applying group-level characteristics to individuals.

Effect modification

Refers to the scenario when the relationship between an exposure and an outcome varies on the basis of a third variable. For instance, perhaps yoga prevents ACL injuries in females but not males. Sex in that scenario is the effect modifier. Effect modification is not the same as *confounding*.

Endemic

The amount of a disease usually found in a given area. Known through *surveillance*.

Epidemic

The occurrence, in a community or region, of cases of an illness (or specific health-related behaviour or other health-related events) clearly in excess of normal expectancy. Epidemiologists and other public health professionals keep track of what levels are "expected" through *surveillance*.

Epidemiology

The study of the distribution and determinants of disease or other health-related events in human populations, and the application of that study to prevent and control health problems.

Etiology

The sum of what is known about how a disease process develops within an individual, including known *determinants.*

External validity

The extent to which we can apply a study's results to other people in the *target population.* Synonymous with *generalizability.* External validity is irrelevant if a study lacks *internal validity.*

Generalizability

See *external validity.*

Gold standard

The best that is currently available. Not necessarily the most feasible.

Incidence

A *measure of disease frequency* that quantifies occurrence of new disease. There are two types, *incidence proportion* and *incidence rate.* Both of these have "number of new cases" as the numerator; both can be referred to as just "incidence." Both must include time in the units, either actual time or person-time. Also called *absolute risk.*

Incidence density

See *incidence rate.*

Incidence proportion

A *measure of disease frequency*. The numerator is "number of new case" and the denominator is "the number of people who were at risk at the start of follow-up." Sometimes if the denominator is unknown, you can substitute the population at the mid-point of follow-up (an example would be the incidence of ovarian cancer in Oregon. We would know how many new cases popped up in a given year, via cancer *surveillance* systems. To estimate the incidence proportion, we could divide by the number of women living in Oregon on July 1 of that year. This of course is only an estimate of the true incidence proportion, as we don't know exactly how many women lived here, nor do we know which of them might not have been at risk of ovarian cancer.) The units for incidence proportion are "per unit time." You can adjust this if necessary (ie if you follow people for 1 month, you can multiply by 12 to estimate the incidence for 1 year). You can (read: should) also adjust the final answer so that it looks "nice." For instance, 13.6/100,000 in 1 year is easier to comprehend than 0.000136 in 1 year. Also called *risk* and *cumulative incidence.*

Incidence rate

A *measure of disease frequency*. The numerator is "number of new cases." The denominator is "sum of the *person-time at risk*." The units for incidence rate are "per person-[time unit]", usually but not always person-years. You can (and should) adjust the final answer so that it looks "nice." For instance, instead of 3.75/297 person-years, write 12.6 per 1000 person-years. Also called *incidence density.*

Incident cases

All new cases of a particular disease, arising over some period of time.

Incubation period

The amount of time between an exposure and the onset of symptoms. Roughly, the *induction period* plus the *latent period.*

Induction period

The amount of time between an exposure and the biological onset of disease. Depending on the exposure/disease pair in question, can vary from minutes for some potent toxins to decades for many chronic diseases.

Internal validity

The extent to which a study's methods are sufficiently correct that we can believe the findings as they apply that that study sample.

Latent period

The amount of time between biological onset of disease and diagnosis. Depending on the disease, can be highly-variable in length, from hours to years. Duration of the latent period also varies depending on access to healthcare.

Measure of association

Quantifies the degree to which a given exposure and outcome are related statistically. Implies nothing about whether the association is causal. Examples of measures of association are *odds ratios, risk ratios, rate ratios, risk differences*, etc.

Measures of disease frequency

Quantifies how much disease is in a population. See *count*, *incidence*, and *prevalence*.

Misclassification bias

Systematic error that results from something (either the exposure, the outcome, a confounder, or all three) having been measured incorrectly. Examples include people not being able to tell you something, people not being willing to tell you something, and an objective measure that is somehow systematically wrong (eg, always off in the same direction, like a blood pressure cuff that is not zeroed correctly). Recall bias, social desirability bias, interviewer bias--these are all examples of misclassification bias. The end result of all of them is that people are put into the wrong box in a 2×2 *table*. If the misclassification is equally distributed between the groups (eg, both exposed and unexposed have equal chance of being put in the wrong box), it's *non-differential misclassification*. Otherwise, it's *differential misclassification*.

Missing at random

All studies have missing data, and many statistical analyses assume that they are missing at random, meaning any given participant is as likely as any other to have missing data. This assumption is almost never met; the kinds of participants who have missing data are usually fundamentally different than those who have more complete data.

Morbidity

Any adverse health outcome short of death.

Mortality

Death.

Negative predictive value (NPV)

One of four test characteristics used to describe the accuracy of screening/diagnostic tests. NPV is the probability that one does not have the disease, given that one tested negative. Calculated as D/(C+D) in standard 2×2 notation (TN/(FN+TN)). Varies as disease prevalence varies.

Non-differential misclassification

Misclassification that occurs equally among all groups.

Null hypothesis

Used in *statistical significance* testing. The null hypothesis is always that there is not difference between the two groups under study.

Null value

The value taken by a *measure of association* if the exposure and disease are not related. Is equal to 1.0 for *relative measures of association*, and equal to 0.0 for *absolute measures of association.*

Observational studies

All study designs in which participants choose their own exposure groups. Includes *cohorts, case-control, cross-sectional.* Basically, includes all designs other than *randomized controlled trial.*

Odds ratio

A *measure of association*, used in study designs that deal with *prevalent cases* of disease (*case-control, cross-sectional*). Calculated as AD/BC, from a standard *2x2 table.* Abbreviated OR.

P-value

A way of quantifying *random error.* The correct interpretation of a p-value is: the probability that, if you repeated the study (go back to the target population, draw a new sample, measure everything, do the analysis), you would find a result at least as extreme, assuming the *null hypothesis* is true. If it's actually true that there's no difference between the groups, but your study found that there were 15% more smokers in group A with a p-value of 0.06, then that means that there's a 6% chance that, if you repeated the study, you'd again find 15% (or a bigger number) more smokers in one of the groups. In public health and clinical research, we usually use a cut-off of $p < 0.05$ to mean "*statistically significant*"--so, we are allowing a *type I error* rate of 5%. Thus, 5% of the time we'll "find" something, even though really there isn't a difference (ie, even though really the null hypothesis is true). The other 95% of the time, we are correctly rejecting the null hypothesis and concluding that there is a difference between the groups.

Period prevalence

Prevalence calculated over a longer period of time. Used for short-duration infectious diseases or injuries.

Person-time at risk (PTAR)

For participants enrolled in a *cohort study* or *randomized controlled trial*, this is the amount of time each person spent at risk of the disease or health outcome. A person stops accumulating person-time at risk (usually shortened to just "person-time") when: (1) they are lost to follow-up; (2) they die (or otherwise not become a risk) of something else other than the disease under study (ie they die of a *competing risk*); (3) they experience the disease or health outcome under study (now they are an *incident case*); or (4) the study ends. Each person enrolled in such a study could accumulate a different amount of person-time at risk.

Point estimate

The *measure of association* that is calculated in a study. Typically presented with a corresponding 95% *confidence interval.*

Point prevalence

Prevalence calculated at a specific moment in time.

Population

A group of people who share a common characteristic.

Population at risk

All individuals in a *population* who (1) have not yet experienced the disease or health outcome under study; and (2) are capable of experiencing that disease or health outcome. In other words, the population at risk excludes all *prevalent cases*, as well as those who for some reason could never experience the outcome (eg, biological males cannot have endometrial cancer). It is not always possible to correctly identify those in the latter group, depending on the disease or health outcome in question. For instance, technically, if we were studying pregnancy, we would need to exclude all women who are either themselves infertile or who are in a monogamous relationship with a man who is infertile. However, in practice it is difficult to identify infertile couples (those who have never tried to get pregnant won't know they're infertile); in such a scenario one would just acknowledge the limitation (that the calculation of population at risk was imperfect, and why).

Positive predictive value (PPV)

One of four test characteristics used to describe the accuracy of screening/diagnostic tests. PPV is the probability that one has the disease, given that one tested positive. Calculated as A/(A+B) or TP/(TP+FP) in standard 2×2 notation. Varies as disease prevalence varies.

Power

The probability that your study will find something that is there. Power = $1 - \beta$; beta is the *type II error* rate. Small studies, or studies of rare events, are typically under-powered.

Prevalence

A *measure of disease frequency* that quantifies existing cases. The numerator is "all cases" and the denominator is "the number of people in the population." Usually expressed as a percent unless the prevalence is quite low, in which case write it as "per 1000" or "per 10,000" or similar. There are no units for prevalence, though it is understood that the number refers to a particular point in time.

Prognosis

The likely course of a disease; how well someone with the disease will fare, given current treatment regimens.

Prophylaxis

Treatment undertaken in an attempt to prevent a poor outcome. It is designed specifically to prevent, not to treat. For instance, in chapter 9, there is discussion of "risk-reducing mastectomy"—prophylactic removal of breasts in women at very high risk of breast cancer. The mastectomy occurs prior to the cancer, in an attempt to prevent the cancer from occurring. As another example, health care workers known to have been exposed to HIV (e.g., from an accidental needle stick) are offered prophylactic anti-retroviral drugs, in an attempt to prevent their bodies from seroconverting/becoming infected with HIV.

Prospective cohort study

See *cohort study*.

Public health surveillance

See *surveillance*.

Publication bias

Bias in the state of the literature on a particular topic that results from journals preferentially publishing papers with exciting results, rather than those showing no effect.

Random error

Inherent in all measurements. "Noise" in the data. Will always be present, but the amount depends on how precise your measurement instruments are. For instance, bathroom scales usually have 0.5 – 1 pound of random error; physics laboratories often contain scales that have only a few micrograms of random error (those are more expensive, and can only weigh small quantities). One can reduce the amount by which random error affects study results by increasing the sample size. This does not eliminate the random error, but rather better allows the researcher to see the data within the noise. Corollary: increasing the sample size will decrease the *p-value*, and narrow the *confidence interval*, since these are ways of quantifying random error.

Randomized controlled trial (RCT)

An intervention (experimental) study. Just like a prospective cohort except that the investigator tells people randomly whether they will be exposed or not. So, grab an at-risk (non-diseased) *sample* from the *target population*, randomly assign half of them to be exposed and half to be non-exposed, then follow looking for *incident cases* of disease. The correct measure of association is the *risk ratio* or *rate ratio*. If done with a large enough sample, RCTs will be free from *confounders* (this is their major strength), because all potential co-variables will be equally distributed between the two groups (thus making it so that no co-variables are associated with the exposure, a necessary criterion for a confounder). Note that the 'random' part is in assigning the exposure, NOT in getting a sample (it does not need to be a 'random sample'). RCTs are often not do-able because of ethical concerns.

Rate ratio

A measure of association calculated for studies that observe *incident cases* of disease (*cohorts* or RCTs). Calculated as the incidence proportion in the exposed over the incidence proportion in the unexposed, or A/(A+B) / C/(C+D), from a standard *2x2 table*. Note that 2x2 tables for cohorts and RCTs show the results at the end of the study--by definition, at the beginning, no one was diseased. See also *rate ratio* and *relative risk*. Abbreviated RR.

Recall bias

A subset of *misclassification bias* that specifically results from people being unable to accurately recall past exposures.

Relative measure of association

A *measure of association* calculated fundamentally by division. See also *risk ratio*, *rate ratio*, *relative risk*, *odds ratio*.

Relative risk

Abbreviated RR. Can refer either to *risk ratio* or *rate ratio*--because of this uncertainty, this term is not used in this book.

Retrospective cohort study

See *cohort study*.

Risk

See *Incidence Proportion.*

Risk difference

A *measure of association* calculated for studies that observe *incident cases* of disease (*cohorts* or RCTs). Calculated as the *incidence proportion* in the exposed minus the incidence proportion in the unexposed.

Risk factors

Variables known to be associated with a disease. May or may not be causally-related.

Risk ratio

A measure of association calculated for studies that observe *incident cases* of disease (*cohorts* or RCTs). Calculated as the incidence proportion in the exposed over the incidence proportion in the unexposed, or A/(A+B) / C/(C+D), from a standard *2x2 table*. Note that 2x2 tables for cohorts and RCTs show the results at the end of the study--by definition, at the beginning, no one was diseased. See also *rate ratio* and *relative risk*. Abbreviated RR.

Sample

The group actually enrolled in a study. Hopefully the sample is sufficiently similar to the target population that we can say something about the *target population*, based on results from our sample. In epidemiology we often don't worry about getting a "random sample"–that's necessary if we're asking about opinions or health behaviours or other things that might vary widely by demographics, but not if we're measuring disease etiology or biology or something else that will likely NOT vary widely by demographics (for instance, the mechanism for developing insulin resistance is likely the same in all humans). Nonetheless, if the sample is different enough than the target population, that is a form of *selection bias*, and can be detrimental in terms of *external validity*.

Screening

Applying a clinical test to asymptomatic individuals, on the theory that finding (and treating) the disease earlier will lead to better outcomes.

Selection bias

A type of systematic error resulting from who chooses/is chosen to be in a study and/or who drops out of a study. Can affect either *internal validity* or *external validity*.

Sensitivity

One of four *test characteristics* used to describe the accuracy of screening/diagnostic tests. Sensitivity is the probability that one tests positive, given that one has the disease. Calculated as A/(A+C) or TP/(TP+FN) in standard 2×2 notation. Does not vary as disease prevalence varies.

Specificity

One of four *test characteristics* used to describe the accuracy of screening/diagnostic tests. Specificity is the probability that one tests negative, given that one does not have the disease. Calculated as D/(B+D) or TN/(TN+FP) in standard 2×2 notation. Does not vary as disease prevalence varies.

Statistical significance

A somewhat-arbitrary method for determining whether or not to believe the results of a study. In clinical and epidemiologic research, statistical significance is typically set at $p < 0.05$, meaning a *type I error* rate of <5%. As with all statistical methods, pertains to random error only; a study can be statistically significant but not believable, eg, if there is likelihood of substantial bias. A study can also be statistically significant (eg, p was < 0.05) but not clinically significant (eg, if the different in systolic blood pressure between the two groups was 2 mm Hg—with a large enough sample this would be statistically significant, but it matters not at all clinically).

Surveillance

The ongoing, systematic collection, analysis, and interpretation of health data, essential to the planning, implementation, and evaluation of public health practice, closely integrated with the timely dissemination to those who need to know. Surveillance both (1) provides information for descriptive epidemiology (person, place, time), and (2) allows us to know what "normal" is, so that potential *epidemics* are identified early. Also called *public health surveillance*.

Target population

The group about which we want to be able to say something. One only very rarely is able to enroll the entire target population into a study (since it would be millions and millions of people), and so instead we draw a *sample*, and do the study with them. In epidemiology we often don't worry about getting a "random sample"--that's necessary if we're asking about opinions or health behaviors or other things that might vary widely by demographics, but not if we're measuring disease etiology or biology or something else that will likely not vary widely by demographics (for instance, the mechanism for developing insulin resistance is the same in all humans).

Test characteristics

Four numerical summaries that describe different aspects of screening/diagnostic test accuracy. Two of the test characteristics (*sensitivity* and *specificity*) are "fixed," meaning their values do not change as disease *prevalence* changes. The other two (*positive predictive value* and *negative predictive value*) do change as disease prevalence changes.

Type I error

The probability that a study "finds" something that isn't there. Typically represented by α, and closely related to *p-values*. Usually set to 0.05 for clinical and epidemiologic studies.

Type II error

The probability that a study did not find something that was there. Typically represented by β, and closely related to *power*. Ideally will be above 90% for clinical and epidemiologic studies, though in practice this often does not happen.

About the Author

Marit L. Bovbjerg, PhD, MS is an assistant professor of epidemiology at Oregon State University, and an advocate of Open Access. As an undergraduate, she studied music and chemistry at the University of Virginia, receiving her Master's degree in Health Evaluation Sciences from that same institution in 2002. She then worked as a clinical research coordinator at the UVa Medical Center for several years before beginning her doctoral training in epidemiology at the University of North Carolina at Chapel Hill. She has been on the faculty at OSU since 2009. She has taught introductory epidemiology to undergraduates, graduate students in public health and related fields, midwives, medical students, and medical residents and fellows.

Dr. Bovbjerg's research program focuses on maternity care in the US, with particular interest in midwifery, community birth, and doula care. She has published numerous peer-reviewed articles, including the 2016 Article of the Year in the *Journal of Midwifery & Women's Health* (on waterbirth) and a recent methods paper questioning the utility of Apgar scores in research, published in the *American Journal of Epidemiology*. Dr. Bovbjerg writes the bi-monthly "Current Resources for Evidence-Based Practice" column for the *Journal of Obstetric, Gynecologic, and Neonatal Nursing*. She has presented her work at numerous national and international conferences. She is the co-director of both the Uplift Research and Health Equity Laboratory and the Community Doula Program. Her work has been funded by HRSA, NIH, the Foundation for the Advancement of Midwifery, and the InterCommunity Health Network, among others. During the 2019-2020 academic year, she was appointed as a Fulbright Scholar to the National Perinatal Epidemiology Centre, University College Cork, Republic of Ireland.

Creative Commons License

Recommended Citations

APA (7th)

Online:

Bovbjerg, M. (2020, October 1). *Foundations of Epidemiology*. https://open.oregonstate.education/epidemiology/.

Print:

Bovbjerg, M. (2020). *Foundations of Epidemiology*. Oregon State University.

APA (6th)

Online:

Bovbjerg, M. (2020, October 01). Foundations of Epidemiology. Retrieved [Retrieval date e.g. January 1, 2021], from https://open.oregonstate.education/epidemiology/

Print:

Bovbjerg, M. (2020). *Foundations of Epidemiology*. Corvallis, OR: Oregon State University.

MLA (8th)

Online:

Bovbjerg, Marit. *Foundations of Epidemiology*. 1 Oct. 2020, open.oregonstate.education/epidemiology/.

Print:

Bovbjerg, Marit. *Foundations of Epidemiology*. Oregon State University, 2020.

MLA (7th)

Online:

Bovbjerg, Marit. "Foundations of Epidemiology." 01 Oct. 2020. Web. [Retrieval date e.g. 1 Jan. 2021].

Print:

Bovbjerg, Marit. *Foundations of Epidemiology*. Corvallis: Oregon State U, 2020. Print.

Chicago

Online:

Bovbjerg, Marit. "Foundations of Epidemiology," October 1, 2020. https://open.oregonstate.education/epidemiology/.

Print:

Bovbjerg, Marit. *Foundations of Epidemiology*. Corvallis, OR: Oregon State University, 2020.

Versioning

This page provides a record of changes made to this publication. Each set of edits is acknowledged with a 0.01 increase in the version number. The exported files, available on the homepage, reflect the most recent version.

If you find an error in this text, please fill out the form at bit.ly/33cz3Q1

Version	Date	Change Made	Location in text
0.1	MM/DD/YYYY		